76th Conference
on Glass Problems

76th Conference on Glass Problems

A Collection of Papers Presented at the 76th Conference on Glass Problems, Greater Columbus Convention Center, Columbus, Ohio, November 2–5, 2015

Editor

S. K. Sundaram

The American Ceramic Society

WILEY

Published by John Wiley & Sons, Inc., Hoboken, New Jersey.
Published simultaneously in Canada.

For general information on our other products and services or for technical support, please contact our Customer Care Department within the United States at (800) 762-2974, outside the United States at (317) 572-3993 or fax (317) 572-4002.

Wiley also publishes its books in a variety of electronic formats. Some content that appears in print may not be available in electronic formats. For more information about Wiley products, visit our web site at www.wiley.com.

Library of Congress Cataloging-in-Publication Data is available.

ISBN: 978-1-119-27499-5
ISBN: 978-1-119-27500-8 (special edition)
ISSN: 0196-6219

Printed in the United States of America.

10 9 8 7 6 5 4 3 2 1

Contents

ENVIRONMENTAL

MODELING

FORMING

Foreword

The 76th Glass Problems Conference (GPC) is organized by the Kazuo Inamori School of Engineering, The New York State College of Ceramics, Alfred University, Alfred, NY 14802, and The Glass Manufacturing Industry Council (GMIC), Westerville, OH 43082. The Program Director was S. K. Sundaram, Inamori Professor of Materials Science and Engineering, Kazuo Inamori School of Engineering, The New York State College of Ceramics, Alfred University, Alfred, NY 14802. The Conference Director was Robert Weisenburger Lipetz, Executive Director, Glass Manufacturing Industry Council (GMIC), Westerville, OH 43082. Donna Banks of the GMIC coordinated the events and provided support. The themes and chairs of five half-day sessions were as follows:

Energy and Glass Melting
Jans Schep—Owens-Illinois, Inc., Perrysburg, OH and Elmer Sperry, Libbey Glass, Toledo, OH

Batching
Phil Tucker, Johns Manville, Denver, CO and Andrew Zamurs, Rio Tinto Minerals, Greenwood, CO, and Martin Goller, Corning Inc.

Combustion, Refractories, and Sensors
Laura Lowe—North American Refractory Company, Pittsburgh, PA and Larry McCloskey—Anchor Acquisition, LLC, Lancaster, OH

Environmental
Glenn Neff, Glass Service, Stuart, FL and James Uhlik, Toledo Engineering Company, Inc., Toledo, OH

Modeling
Bruno Purnode, Owens Corning Composite Solutions, Granville, OH and Warren Curtis, PPG Industries, Pittsburgh, PA

Forming

Kenneth Bratton, Emhart Glass Research Inc., Windsor, CT and Uyi Iyoha, Praxair Inc., Tonawanda, NY

Preface

This volume is a collection of papers presented at the 76th year of the Glass Problems Conference (GPC) in 2015. This conference continues the tradition of publishing the papers that dates back to 1934. The manuscripts included in this volume are reproduced as furnished by the presenting authors, but were reviewed prior to the presentation and submission by the respective session chairs. These chairs are also the members of the GPC Advisory Board. I appreciate all the assistance and support by the Board members. The American Ceramic Society and myself did minor editing and formatting of these papers. Neither Alfred University nor GMIC is responsible for the statements and opinions expressed in this volume.

As the Program Director of the GPC, I enjoy interacting with industry experts, representatives, and students at the GPC. I am thankful to all the presenters at the 76th GPC and the authors of these papers. The GPC continues to grow stronger with the support of the teamwork and audience. I appreciate all the support from the members of Advisory Board. Their volunteering sprit, generosity, professionalism, and commitment were critical to the high quality technical program at this Conference. I also appreciate the continuing support and leadership from the Conference Director, Mr. Robert Weisenburger Lipetz, Executive Director of GMIC and excellent support from Ms. Donna Banks of GMIC in organizing the GPC. I look forward to continuing our work with the entire team in the future.

S. K. Sundaram
Alfred, NY
December 2015

Acknowledgments

It is a great pleasure to acknowledge the dedicated service, advice, and team spirit of the members of the Glass Problems Conference (GPC) Advisory Board (AB) in planning this Conference, inviting key speakers, reviewing technical presentations, chairing technical sessions, and reviewing manuscripts for this publication:

Kenneth Bratton—*Emhart Glass Research Inc. Hartford, CT*
Warren Curtis—*PPG Industries, Inc., Pittsburgh, PA*
Martin Goller—*Corning Incorporated, Corning, NY*
Uyi Iyoha—*Praxair Inc.,Tonawanda, NY*
Robert Lipetz—*Glass Manufacturing Industry Council, Westerville, OH*
Laura Lowe—*North American Refractory Company, Pittsburgh, PA*
Larry McCloskey—*Anchor Acquisition, LLC, Lancaster, OH*
Glenn Neff—*Glass Service USA, Inc., Stuart, FL*
Adam Polcyn*—*PPG Industries, Inc., Pittsburgh, PA*
Bruno Purnode—*Owens Corning Composite Solutions, Granville, OH*
Jans Schep—*Owens-Illinois, Inc., Perrysburg, OH*
Elmer Sperry—*Libbey Glass, Toledo, OH*
Phillip Tucker—*Johns Manville, Denver, CO*
James Uhlik—*Toledo Engineering Co., Inc., Toledo, OH*
Justin Wang*—*Guardian Industries Corporation, Geneva, NY*
Andrew Zamurs—*Rio Tinto Minerals, Greenwood, CO*

In addition, I am indebted to Donna Banks, GMIC for her patience, support, and attention to detail in making this conference a success.

*Joined the AB at the 76th GPC.

Energy and Glass Melting

STRENGTH OF GLASS AND GLASS FIBERS

Hong Li
Fiber Glass Science and Technology, PPG Industries, Inc.
Pittsburgh, Pennsylvania, USA

ABSTRACT

The article provides a selective review on strength of glass and glass fiber, covering effects of surface flaw and surface hydrolysis on the usable strength of glass (USG). Application of Griffith-Inglis-Orowan theory on fracture of solids is demonstrated, elucidating importance of stress-assisted hydrolytic effect on glass USG and associated change in glass surface energy. The fundamental understanding of glass fracture supports critical needs for development of new glasses and new durable and/or resin compatible hydrophobic coatings to significantly improve USG of glass and fiberglass products, respectively.

1. FRACTURE OF GLASS AND GLASS FIBERS

1.1 Fundamental of Solid Fracture

Theoretical tensile strength of solids, according to Orowan [1], is proportional to Young's modulus (E) and surface energy (γ_o) of the material as

$$\sigma_{th} = (E\gamma_o/r_o)^{1/2} \tag{1}$$

where r_o is the equilibrium distance between atomic centers. Experimental measurements, however, report that glasses typically have tensile strengths much lower than the theoretical values by as much as one order of magnitude. Unlike crystalline materials, for which grain boundaries serve as one type of defect, glass defects mostly come from surface "damage" or surface flaw as one of the key factors of lowering the usable strength of glass (USG) from its expected theoretical level.

Surface flaws of a given size (c) serve as a stress concentrator when glass is subject to an applied tensile load; these weak spots cause glass to fail at a tensile stress level well below the theoretical expectation. By the Griffith energy-balance criterion, apparent or measured strength (σ_m) of a solid is defined by [2, 3]:

$$\sigma_m = (2E\gamma_o/\pi c)^{1/2} \quad \text{(plane tensile stress)} \tag{2a}$$

$$\sigma_m = [2E\gamma_o/\pi(1-v^2)c]^{1/2} \quad \text{(plane tensile strain)} \tag{2b}$$

Inglis further demonstrated [4] that tip geometry of the flaw, in terms of its size c and radius, ζ_{tip}, can significantly magnify the stress applied onto the material, which affects σ_m, according to

$$\sigma_m = (E\gamma_o/4r_o)^{1/2} (\zeta_{tip}/c)^{1/2} \tag{3a}$$

or $\quad \sigma_m = \sigma_{th}/2 (\zeta_{tip}/c)^{1/2} \tag{3b}$

Equation 3b implies that the maximum measured strength of "flaw-free" samples will be approximately 50% of its theoretical strength and the same size of a critical surface flaw with a sharper crack tip (or lower radius at the crack tip) will further reduce the material strength [5].

It becomes clear that experimentally measured glass strength is not an intrinsic property of the material. Besides composition, atomic structures of glass are affected by their thermal history in terms of melting temperature, cooling rate, degree of annealing, degree of aging under conditions under which they are stored before application, and fatigue in terms of test or application conditions, including temperature, humidity, and cleanness of laboratory, and sample strain rate [6-9]. Furthermore, it is expected that the glass "surface defects" can be generated from "contact damage" even from finger contact during sample handling.

When developing new glass and glass fiber compositions, keeping in mind the multiple factors that affect USG, it is critical to test all samples that are made by the same method under the same laboratory conditions in order to screen composition effect on glass strength.

In reporting and comparing glass strength, "pristine" strength refers to testing samples made under controlled humidity, not being "damaged" by any physical contacts in handling, and tested under the same humidity environment within a very short period of time after the samples are made. "Inert" strength means that the samples are tested in liquid nitrogen to minimize any moisture interaction with glass or glass fiber surface under an applied force. In this case, the samples can be tested after aging under specific conditions or as its "pristine" form without any treatment. "Inert" strength of the "pristine" glass is significantly higher than that of "pristine" glass and hence, closer to the glass intrinsic property.

1.2 Glass Fracture from Microscopic Defects

One of the most detrimental factors impacting glass strength is glass surface attack by corrosive media in the form of liquid or vapor, including water, acid, and base [10-14]. Figure 1 illustrates the effect of fiber surface flaw geometry on silica fiber "inert" tensile strength as the fibers treated in hydrofluoric acid vapor over time [5]. Prediction from the data set suggests that for the silica fibers with very sharp surface flaws, i.e. $\zeta_{tip} \ll c$, its strength is approximately 35 - 40% of its theoretically predicted value of ≥ 17 GPa.

Figure 2 shows fiber failure strain of boron-free E-CR fibers with and without aging up to 270 days at 50°C under 80% relative humidity (RH) [15]. The tests were conducted by using the two-point bending method [16] at room temperature (RT) under 50 %RH and in liquid nitrogen (LN), respectively. Several characteristics can be summarized from the results as follows: First, at semi-logarithmic scale, the two sets of data can be reasonably fit by using linear aggression least square method. The total reduction in fiber failure strain is about 12.5% for fibers tested at RT – 50%RH and 13% for fiber tested in liquid nitrogen, respectively. Therefore, it is reasonable to conclude that fiber aging under stress-free conditions results in approximately 13% deterioration in terms of failure strain. Secondly, in terms of absolute failure strain comparing the two test conditions, ε_f (LN) is significantly higher than ε_f (RT-50%RH); the ratio of the average values for the same aging durations between the two cases is between 2.2 and 2.3, supporting that fiber failure at much higher load or applied stress once moisture of water is minimized or eliminated under which the samples are tested.

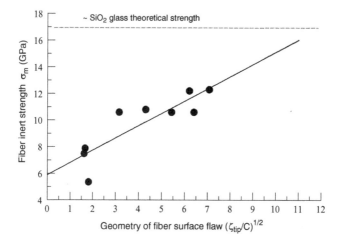

Figure 1. Silica Glass Fiber Tensile Strength as a Function of Fiber Surface Defect Geometry Characterized by the Ratio of Tip Radius (ζ_{tip}) of the Surface Flaw over the Flaw size (c) (solid line is determined by using least square linear regression analysis; the plot is constructed based on [5]).

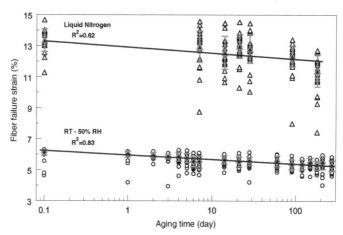

Figure 2. Fiber Failure Strain of E-CR Fibers Measured at Room Temperature under 50% RH and in Liquid Nitrogen as a Function of Fiber Aging under 50°C - 80% RH Conditions (open circle and triangle represent individual measurements; filled diamond and triangle represent average values; error bars represent one standard deviation; solid lines are obtained by using linear regression method fitting average values of the data sets; 20 measurements were performed per data set) [15].

Combining the aging test results, for E-CR glass with low alkali contents and free from boron and fluoride, the study demonstrated that moisture water interaction with the surfaces of fibers being under tensile strain or tensile stress plays a dominant role on its failure over hydration or aging treatment of the fibers without being stressed. It follows that the kinetics of stress-assisted hydrolysis on the fiber surfaces should not be significantly affected by the preexisting "layer" of hydration created from the aging treatment. In turn, one can reason that the hydroxyl groups (Si-OH) formed on the fiber surfaces during aging should be immobile during the growth of crack under the applied stress or strain and hence, newly generated Si-OH groups at the front of surface flaws, i.e., stress-assisted hydrolysis, should dominate the fiber failure strain or failure stress. The mechanism of glass fatigue in a humid environment was proposed and experimentally demonstrated by Hillig & Charles [17] and Wiederhorn [18, 19].

The stress-assisted hydrolysis of the glass near the tip of surface flaws can result in significant glass surface energy (γ) reduction; literature data shows that quartz crystals change in surface energy with and without hydration by as much as 10 times [20-23]. Structure of crystalline quartz and fused quartz glass is very different; in dry liquid nitrogen their perspective ratio is about 0.43 (2.0 J/m^2 for crystalline quartz over 4.6 J/m^2 for fused quartz glass) [24]. However, their perspective changes in surface energy to hydration are expected to follow the same trend [25]. Therefore, the glass fibers tested should become much weaker under ambient conditions over liquid nitrogen. Our estimation on the surface energy ratio, γ(LN)/γ(RT-50%RH), derived from the study [15] was close to 3.4 $\pm$ 0.2 for fibers aged up to 180 days. The surface energy ratio can be derived from Eq. 3a, in which fiber modulus is considered as a strain-dependent variable, i.e., Secant Modulus, according to Gupta and Kurkjian [26].

1.3 Glass Fracture from Macroscopic Defects

As the size of glass surface flaws becomes larger, glasses fail at lower applied stresses, i.e., lower USG, as illustrated in Figure 3 [27]. Within each flaw size range, instantaneous strength represents the samples without any aging effect, and endurance limit represents the samples experienced some levels of aging event before or during the mechanical tests. In product design, one should consider the use of the endurance limit of the glass that has been tested under a relevant application environment to ensure the maximum safety of the products to be used.

Fracture of glass and glass fibers under an applied tensile load initiates at a point of its weakest point according to the Weibull statistical theory [28], which has been widely used to study distribution of glass stress at breakage in relationship to change of glass chemistry or glass thermal history or test conditions. According to the Weibull method, an accumulative probability of failure (P_f) of a solid at an applied tensile stress, σ_f, follows

$$P_f = 1 - \exp[-(\sigma_f/\sigma_o)^\beta] \tag{4a}$$

$$\text{or} \quad \ln[-\ln(1-P_f)] = \beta\ln(\sigma_f) - \beta\ln(\sigma_o) \tag{4b}$$

where β and σ_o are the statistical linear regression fitting parameters, which are often called Weibull modulus (or shape parameter) and characteristic stress, respectively.

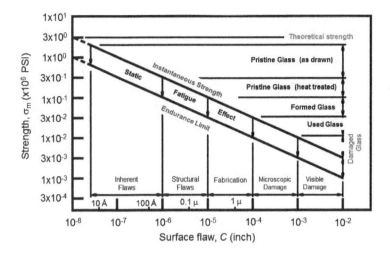

Figure 3. Effect of Surface Flaw Size on Glass Tensile Strength (redraw after Mould [27])

Figure 4 illustrates a Weibull plot of pristine tensile strength distributions of S-Glass, R-Glass, and E-Glass fibers. Under the same sample preparation and test conditions, the average fiber tensile strength ranks in an order S-Glass (5500±133 MPa) > R-Glass (4135±280 MPa) > E-Glass (3215±198 MPa). The Weibull modulus (β-value) of the S-Glass is substantially higher than both R-Glass and E-Glass [29]. In this case, to minimize the size effect on the fiber strength [6,7], the fiber gage length of all samples was kept the same (1 inch) and the diameter of the fibers was controlled at 10±0.5 μm; the size can be attributed to the change in the defect population as fiber gage length and/or diameter varies.

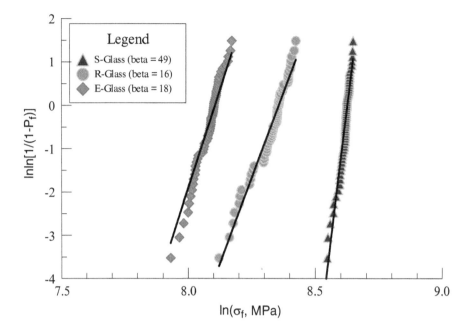

Figure 4. Weibull Plot of Representative Glass Fibers used for Plastic Reinforcement, E-Glass, R-Glass, and S-Glass [29].

2. YOUNG'S MODULUS OF GLASS AND GLASS FIBERS

The strength of glass is a function of Young's modulus (Eq.1 & 2). For oxide glasses, Sun's binding energy approach [30] has been adopted to calculate Young's modulus by Makishima and Mackenzie [31]. The model was later modified by Zou and Toratani [32]. In both models Young's modulus of glass is approximated by a linear combination of contributions from individual glass constituents. Similar approaches to predicting the Young's modulus of complex glass systems can be also found elsewhere [33, 34]. A general presentation of a linear composition model is illustrated in Figure 5, which provides a simplified view of the listed oxide contributions to glass Young's modulus.

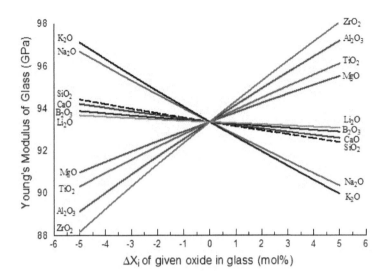

Figure 5. Oxide Effects on Silicate Glass Young's Modulus [29]

In practice, significant deviations between the measured and the model-derived values have been reported, especially in complex multi-component glass systems [35]. There are many key factors contributing to the discrepancies reported. First, local structure or surrounding oxygen environments of glass network formers (SiO_2, B_2O_3) and conditional network formers (Al_2O_3) vary depending on concentrations of alkalis (Li_2O, Na_2O, K_2O), alkaline earth (MgO, CaO, SrO), and their relative proportions [36-41]. The linear composition models cannot account for the structural variations or speciation of the network formers, such as SiO_2, B_2O_3, and Al_2O_3.

Secondly, glass density or molar volume is affected by fictive temperature or thermal history of the samples in terms of glass structure relaxation [42-44]. In turn, annealed glass has lower fictive temperature, higher density, and hence, higher Young's modulus as compared to the quenched form of the same glass composition. Figure 6 compares the measured fiber glass modulus as obtained by using a sonic method comparing with the calculated modulus. In general, a parallel downshift correlation line can be drawn relative to the ideal 1:1 line, which likely results primarily from thermal effect. The glass models of Young's modulus found in literature are built from experimental data generated from testing annealed bulk glass samples [32-34] and the measured values shown in Figure 6 were collected from measuring fast quenching fibers without annealing. The thermally induced change of glass Young's modulus has been reported to vary between 10% and 20% [43, 44].

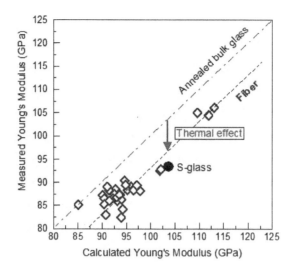

Figure 6. Comparison between Measured Young's Modulus of Glass Fiber and Model Calculated Modulus, Suggesting Thermal Effect.

3. IMPROVEMENT OF USABLE STRENGTH OF GLASS AND GLASS FIBERS

The detrimental effect of surface contact damage on USG has been well understood as reviewed previously, surface protection on glass and glass fibers are important to raise USG. For example, by applying protective sizing on glass fiber, the fiber surface flaw distributions can be altered, resulting in USG improvement [7]. The second critical damage mechanism is glass surface hydrolysis under the applied load. The hydrolysis involves interaction of moisture water or other corrosive media with surfaces of glass or glass fibers. For alkali-free glasses, the adverse effect of hydrolysis on USG is mostly pronounced when the glass object is stressed as comparing with the glass object aged under the stress-free conditions (cf. Fig 2).

Realizing the importance of stress-assisted glass hydrolysis effect, it becomes clear that it will be advantageous by applying a hydrophobic coating on bulk glass [45] or more hydrophobic, resin compatible sizing on glass fibers [48, 49], which will prevent water molecules access to surface defects or surfaces of glass or glass fibers while the objects are under applied load. Prevention and/or slowing down molecular water reaction with glass surface defects in application can be significant in boosting USG according to the glass fatigue mechanism [17-19]. For example, hermetic coating has been long recognized as an important technology and applied to optical silica fibers in telecommunication, which prevents the optical fibers from aging in field applications [48].

Besides thermal tempering [49], improvement of USG can be realized by using various chemical tempering processes, introducing a compressive layer on surfaces of glass and/or glass fibers depending on specific compositions of the glasses commonly containing appreciable amount of sodium. Two methods have been commonly used in commercial glass productions are glass surface de-alkalization and ion exchange [50-53]. Table 1 summarizes improvement of glass strength after surface treatments with various chemical methods [50].

It should be noted that most of the aforementioned methods, besides the use of sizing, are technically feasible on bulk glasses in form of plate, tubing, and containers; but not feasible in production of continuous fibers.

Table 1. Improvement of Average Impact Resistance (N·cm) of Glasses Treated by using Various Chemical Tempering Processes (De-alkalization, Ion-exchange, and Combined Methods) [50]

De-alkalization Agent	As-received	De-alkalization	Ion-Exchange*	Combined#
SO_2	76	93 (22% ↑)	105.5 (39% ↑)	144 (89% ↑)
NH_4Cl	76	96 (26% ↑)	105.5 (39% ↑)	152 (100% ↑)
$(NH_4)_2SO_4$	76	92 (21% ↑)	105.5 (39% ↑)	135 (78% ↑)
$AlCl_3$	76	88 (16% ↑)	105.5 (39% ↑)	120 (58% ↑)
$(NH_4)_2SO_4+AlCl_3$ (10:1)	76	88 (16% ↑)	105.5 (39% ↑)	130 (71% ↑)
$NH_4Cl+AlCl_3$ (10:1)	76	96 (26% ↑)	105.5 (39% ↑)	129 (70% ↑)
$NH_4Cl+(NH_4)_2SO_4$ (1:1)	76	98 (29% ↑)	105.5(39% ↑)	132 (74%↑)

* Ion-exchange process: soak container in 70°C - 200 ml water solution containing 34g KNO_3 - 69g KCl - 8.5g K_2SO_4 and followed by heat-treatment at 500°C. # De-alkalization first and followed by ion-exchange treatment

4. SUMMARY

A selective literature review on strength of glass and glass fiber was made covering effects of surface flaw and surface hydrolysis on USG. Based on fracture mechanism of Griffith-Inglis-Orowan theory, specific examples are provided to elucidate importance of stress-assist hydrolytic effect on USG, which highlights more pronounced detrimental impact of stress-assisted glass surface hydrolysis over the effect of stress-free hydrolysis. Although it is important to develop new glass chemistry for greater pristine strength as shown in the review, in commercial applications, development of new coating materials for bulk glass or sizing for fiber glass is equally important to raise USG. Especially the latter offers the improvement of USG for existing glass or fiber glass products with minimum or without changing of existing processes, i.e., glass melting and product forming.

ACKNOWLEDGEMENT

The author greatly appreciates Fiber Glass, PPG Industries, Inc. for supporting various fiber glass research projects, based on which some of the results were generated and used in the article and is grateful to several stimulating discussions with R.K. Brow, C.R. Kurkjian, and L. Huang on the related subject.

REFERENCES

[1] E. Orowan, "The Fatigue of Glass under Stress," Nature 154 341-343 (1944).

[2] A.A. Griffith, "The Phenomena of Rupture and Flow in Solids," Philos. Trans. Roy. Soc. Lond. A221 163 – 198 (1920).

[3] A.A. Griffith, "The Theory of Rupture," in Proc. 1st International Congress on Appl. Mech., ed. C.B. Biezeno and J.M. Burgers, (1924) p. 55.

[4] C.E. Inglis, "Stresses in a Plate due to the Presence of Cracks and Sharp Corners," Trans Inst. Naval Archit, 55 219 (1913).

[5] A. Zhong, D. Inniss, C.R. Kurkjian, "Strength-Flaw Relationship of Corroded Pristine Silica Studied by Atomic Force Microscopy," Mat. Res. Soc. Symp. Proc. 322 537-542 (1994).

[6] K.L. Loewenstein and J. Dowd, "An Investigation of the Relationship between Glass Fibre Tensile Strength, the Temperature of the Glass from which the Fibre is Drawn, and Fibre Diameter," *Glass Tech.* 9[6] 164-171 (1968).

[7] L. Yang and J.L. Thomason, "Effect of Silane Coupling Agent on Mechanical Performance of Glass Fibre," J. Mater. Sci. 48 1947-1954 (2012).

[8] 50. V.G. Pähler and R. Brückner, "Strength of Glass Fibres as a Function of Drawing Parameters,"Glastechn. Ber. 54[3] 52-64 (1981).

[9] 51. W.F. Thomas, "An Investigation for the Factors Likely to Affect the Strength and Properties of Glass Fibers," Phys. Chem. Glasses 1 4-18 (1960).

[10] A. Zhong, D. Inniss, C.R. Kurkjian, "Strength-Flaw Relationship of Corroded Pristine Silica Studied by Atomic Force Microscopy," Mat. Res. Soc. Symp. Proc. 322 537-542 (1994).

[11] N.M. Cameron, "The Effect of Environment and Temperature on the Strength of E-glass Fibres. Part 1. High Vacuum and Low Temperature," Glass Tech. 9[1] 14-21 (1968).

[12] N.M. Cameron, "The Effect of Environment and Temperature on the Strength of E-glass Fibres. Part 2. Heating and Ageing," Glass Tech. 9[5] 121-130 (1968).

[13] D.R. Cockram, "Strength of E Glass in Solutions of Different pH," Glass Tech. 22[5] 211-214 (1981).

[14] C.R. Kurkjian, P.K. Gupta, R.K. Brow, and M. Lower, "The Intrinsic Strength and Fatigue of Oxide Glasses," J. Non-Cryst. Solids, 316 114-124 (2003).

[15] Q. Wang, R.K. Brow, H. Li, E.A. Ronchetto, "Effect of Aging on the Failure Characteristics of E Glass Fibers," J. Mater. Sci., November 2015 (DOI 10.1007/s10853-015-9549-0)

[16] R.K. Brow, N.P. Lower, and C.R. Kurkjian. "Two-Point Bend Studies of Glass Fibers- New Insights to Glass Strength and Quality," Ceramic Bulletin, 84[10] 50-54 (2005).

[17] W. B. Hillig and R. J. Charles, "Surface, Stress-Dependent Surface Reactions and Strength," in High Strength Materials, Ed. V. F. Zackey, John Wiley and Sons, inc., New York (1965).

[18] S.M. Wiederhorn "Influence of Water Vapor on Crack Propagation in Soda-Lime Glass," J. Am. Ceram. Soc. 50 [8] 407-14 (1967).

[19] S.M. Wiederhorn and L.H. Bolz, "Stress Corrosion and Static Fatigue of Glass", J. Am. Ceram. Soc., 53 [10] 543-48 (1970)

[20] J.W. Whalen, "Thermodynamic Properties of Water Adsorbed on Quartz," J. Phys. Chem. 65 1676-1681 (1961).

[21] J.W. Whalen, "Heats of Immersion in Silica-Water Systems," Adv. Chem. Ser. 33 281-290 (1961).

[22] G.A. Parks, "Surface and Interfacial Free Energies of Quartz," J. Geophy. Res. 89 [B6] 3997-4008 (1984).

[23] A.K. Kronenberg, "Hydrogen Speciation and Chemical Weakening of Quartz," in Silica, ed. O.H. Ribbe, Rev Mineral. Vol. 29, 124 – 176 (1994).

[24] S.M. Wiederhorn, "Fracture Surface Energy of Glass," J. Am. Ceram. Soc., 52[2] 99-105 (1969).

[25] Yu. K. Shchipalov, "Surface Energy of Crystalline and Vitreous Silica," Glass Ceram. 57 [11-12] 374 – 377 (2000).

[26] P.K. Gupta, C.R. Kurkjian, "Intrinsic failure and non-linear elastic behavior of glasses," J. Non-Cryst. Solids, 351 2324-2328 (2005).

[27] R.E. Mould, "The Strength of Inorganic Glasses," in Fundamental Phenomena in the Materials Sciences, ed. L.J. Bonis, J.J. Duga and J.J. Gilman, 119-149 (1967)

[28] W. Weibull, "A Statistical Distribution Function of Wide Applicability," J. Appl. Mech. 18 293 (1951).

[29] H. Li, C. Richards, J. Watson, "High Performance Glass Fiber Development for Composite Applications." Int. J. Appl. Glass Sci. 5 65-81 (2014).

[30] K.H. Sun, "Fundamental Condition of Glass Formation," J. Am. Cer. Soc. 30 277 - 281 (1947).

[31] A. Makishima and J.D. Mackenzie, "Direct Calculation of Young's Modulus of Glass," J. Non-Cryst. Sol. 12 35-45 (1973).

[32] Z. Zou and H. Toratani, "Compositional Design of High Modulus Glasses for Disk Substrates," J. Non-Cryst. Sol. 290 180-188 (2001).

[33] M.B. Volf, Ch 15, Method of Appen, Mathematical Approach to Glass, Elsevier, New York, 1988, p.143-153.

[34] J. Rocherulle, C. Ecolivet, M. Poulain, P. Verdier, and Y. Laurent, "Elastic Moduli of Oxynitride Glasses, Extension of Makishima and Machenzie's theory," J. Non-Cryst. Sol. 108 187-193 (1989).

[35] R. Jindal, W. Jatmiko, I.V. Singh, and R. Jayaganthan, "Elastic Properties of Clinopyroxene Based Glasses along Diopside ($CaMgSi_2O_6$) – Jadeite ($NaAlSi_2O_6$) Join, "J. Miner. Mater. Charact. Eng. 11[3] 267-283 (2012).

[36] L. Han and S. Zhao, "Development of High Strength and High Modulus Glass Fibers," Fiber Glass 3 34-38 (2011).

[37] S.K. Lee and J.F. Stebbins, "The Distribution of Sodium Ions in Aluminosilicate Glasses: A High-Field Na-23 MAS and 3Q MAS NMR Study," Geochim. Cosmochim. Acta 67[9] 1699-1709 (2003).

[38] S. Wang and J.F. Stebbins, "Multiple-Quantum Magic-Angle Spinning 17O NMR Studies of Borate, Borosilicate, and Boroaluminate Glasses," J. Am. Ceram. Soc., 82[6] 1519-1528 (1999).

[39] H. Yamashita, K. Inoue, T. Nakajin, H. Inoue, and T. Maekawa, "Nuclear Magnetic Resonance Studies of $0.139MO$ (or M'_2O)$\cdot0.673SiO_2\cdot(0.188-x)Al_2O_3 \cdot xB_2O_3$ (M=Mg, Ca, Sr, and Ba, M' = Na and K) Glasses," J. Non-Cryst. Solids, 331 (2003) 128-136.

[40] L.S. Du and J.F. Stebbins, "Network Connectivity in Aluminoborosilicate Glasses: A High-Resolution ^{11}B, ^{27}Al and ^{17}O NMR Study," J. Non-Cryst. Solids, 351 3508-3520 (2005).

[41] F. Angeli, O. Villain, S. Schuller, T. Charpentier, D. Ligny, L. Bressel, and L. Wondraczek," Effect of Temperature and Thermal History on Borosilicate Glass Structure," Phys. Rev. B. 85 054110-1 - 054110-5 (2012)

[42] W.H. Otto, "Compaction Effects in Glass Fibers," J. Am. Ceram. Soc. 44[2] 68-72 (1961).

[43] L. Yang and J.L. Thomason, "The Thermal Behavior of Glass Fibre Investigated by Thermomechanical Analysis," J. Mater. Sci. 48 5768-5775 (2013).

[44] G. Gavriliu, "Effects of Added Oxide and Thermal History on Young's Modulus of Added Oxides(s)-Na_2O-SiO_2 glass," Mater. Lett. 48 199-204 (2001).

[45] C.E. Johnson, D.C. Harris, J.G. Nelson, C.F. Kline, and B.L. Corley, "Strengthening of Glass and Pyroceram with Hydrophobic Coatings," Report NAWCWD TP 8536, Naval Air Warfare Center Weapons Division, China Lake, CA (July 2003).

[46] J.G. Iglesias, J. González-Benito, A.J. Aznar, J. Bravo, J. Baselga, "Effect of Glass Fiber Surface Treatments on Mechanical Strength of Epoxy Based Composite Materials," J. Colloid Inter. Sci. 250[1] 251-260 (2002).

[47] E.N. Brown, A.K. Davis1, K.D. Jonnalagadda, N.R. Sottos, "Effect of Surface Treatment on the Hydrolytic Stability of E-Glass Fiber Bundle Tensile Strength," Comp. Sci. Tech. 65 [1] 129-136 (2005).

[48] C.R. Kurkjian, J.T. Krause, M.J. Matthewson, "Strength and Fatigue of Silica Optical Fibers," J. Lighwave Technology, 7[9] 1360-1370 (1989).

[49] R. Gardon, Ch 5 Thermal Tempering of Glass in Elasticity and Strength in Glasses, ed. D.R. Uhlmann and N.J. Kreidl, Glass Sci. Tech. Vol. 5, Academic Press (New York, 1980) p. 145 - 216

[50] C.Y. Wang and Y. Tao, in Ch 9. Glass Surface Chemical Treatment in Glass Surface Treatment Technology, Chemical Industry Press (Beijing, 2004) p. 222 – 275 (in Chinese).

[51] A.K. Varshneya, W.C. LaCourse, Technology of Ion Exchange Strengthen of Glass: A Review", Ceram. Trans. 29 365-378 (1993).

[52] R. Gy, "Ion Exchange for Glass Strengthening," Mater. Sci. Eng. B149 159-165 (2008).

[53] D.J. Green, "Compressive Surface Strengthening of Brittle Materials by a Residual Stress Distribution" J. Am. Ceram. Soc. 66(11) 807-810 (1983).

OPERATING EXPERIENCE WITH OPTIMELT™ REGENERATIVE THERMO-CHEMICAL
HEAT RECOVERY FOR OXY-FUEL FIRED GLASS FURNACES

A. Gonzalez and E. Solorzano
Grupo Pavisa, S.A. de C.V., Naucalpan, Edo. México
S. Laux, U. Iyoha, K.T. Wu, and H. Kobayashi
Praxair, Inc., Danbury, CT, USA

ABSTRACT
The operation of glass furnaces with oxy-fuel combustion in combination with advanced heat recovery is a compelling low-cost solution. Praxair has developed and demonstrated a regenerative heat recovery system for oxy-fuel fired furnaces that use regenerators in a similar way to which conventional regenerators are used for air preheating. The OPTIMELT™ Thermo-Chemical Regenerator (TCR) technology stores waste heat from the hot flue gas and uses this energy to reform a mixture of natural gas and recirculated flue gas to hot syngas. The natural gas reacts endothermically in the hot checker pack with the water vapor and CO_2 in the recycled flue gas, forming H_2 and CO as a hot syngas fuel and resulting in efficient thermo-chemical heat recovery.

This novel technology for heat recovery was successfully installed on a 50 t/d commercial container glass furnace in the summer of 2014 and has been operated in daily production since then. In addition to a positive impact on production and quality, the TCR system successfully reduced natural gas and oxygen consumption by 15 to 18% relative to the oxy-fuel baselines depending on glass type and cullet rate. This paper summarizes the extensive operating experience with the installation and includes operational data, glass quality, and emission results.

INTRODUCTION
The OPTIMELT™ thermochemical regenerator (TCR) process is an advanced heat recovery technology for oxy fuel fired glass furnaces. The technology utilizes conventional regenerators and endothermic reforming reactions between fuel and recycled flue gas (RFG) to recover flue gas exhaust heat. For a larger scale commercial furnace, expected fuel savings are about 20% compared to oxy fuel and 30% compared to air-regenerator furnaces. A number of papers [1-4] have already been published to introduce this energy savings technology to the glass industry. The technology heats and reforms fuel and RFG mixture in a hot checker pack of the regenerator without catalysts or separate steam generation. The reformed fuel containing hydrogen (H_2) and carbon monoxide (CO) and soot was shown to form a highly luminous flame and to transfer heat efficiently to the glass melt. The regenerators are similar in design to conventional air heating regenerators, but only 1/3 the size in the checker volume, making the retrofit or rebuild an economically attractive option. The TCR system is integrated into the oxy-fuel combustion system and the furnaces can be operated either in oxy-fuel firing, or TCR firing mode.

OPERATING RESULTS AT PAVISA
The TCR system has been in operation on Pavisa's Furnace 13 since September 2014. This furnace produces a wide variety of glass types and colors with frequent product changeovers involving occasional furnace draining. For instance, during a ten-month period between 1 September 2014 and 30 June 2015, the furnace underwent thirteen product changes, which equates to an average production time of about three weeks per glass type. Cullet ratios also fluctuated widely between 25 to 70% and the furnace pull rate varied from 40 to about 54

tpd, depending on the type and the color of the glass produced. It is important to point out that the TCR system has been operating without any loss of glass production or quality issues since completion of the startup in October 2015.

When the furnace was retrofitted with the TCR system the original oxy-fuel combustion system was left operational as a backup. It is also used during times when the furnace is drained during color changes. The original natural draft oxy-fuel flue stack is closed with a refractory damper during TCR operation. The transitions between oxy-fuel and TCR operation are fully automated and only require the operator to initiate the process on the HMI control screen. To start the TCR system from oxy-fuel combustion the control system gradually preheats the regenerators and closes the stack damper to divert more flue gas into the regenerators. At a certain temperature the oxy-fuel burners are shut down while the fuel, oxygen, and RFG flows to the syngas burner and regenerators are ramped up.

In order to better quantify fuel savings with TCR, tests were conducted in the first half of 2015 to compare the fuel consumption under oxy-fuel firing and TCR firing for different glass types. The data selection process focused strictly on operations where the glass type and cullet ratios were the same and the pull rates were closely similar so that the comparisons were well defined without using any major assumptions. Based on the analysis of the longer term (six-month) data, it was confirmed that the TCR system at Pavisa resulted in 15 to 18% fuel savings, compared to the oxy-fuel firing baselines. One factor which limits more fuel savings at Pavisa was the higher relative wall heat losses due to the small furnace size and the two thermochemical regenerators. The larger regenerator wall losses were the results of thinner wall insulations that were imposed by the limited space available near the furnace for the regenerator installations. However, after taking into account all potential losses the results are well within the expectations for the technology at the small furnace scale. The projected fuel savings on larger furnaces is discussed below.

The TCR system was shut down for inspection and scheduled yearly maintenance in July 2015. The access doors on the regenerators were opened and the condition of refractory and walls assessed. The ducts of the flue gas recirculation skid were cleaned and inspected for corrosion. The following is a high-level summary of the findings:

- The checkers were in excellent condition. The individual channels of the checker pack were completely free of obstructions. The individual checker pieces did not show any signs of corrosion. No deposits were visible at the top and only light deposits that were easily removed at the bottom. Figure 1 below shows a picture of the left regenerator top after the opening of the access doors.
- The refractory of the port neck walls showed some surface spalling in the hottest zone. This is likely a result of the frequent cool-downs and heat-ups during startup and optimization as it was observed earlier in the operation of the TCR system. The choice of the refractory is being reviewed and a repair plan is in preparation should it become necessary to intervene.
- All other refractory and the rider arches showed no sign of deterioration.
- The flue gas recirculation system had internal deposits that could be easily cleaned. The duct material and dampers were in very good condition. Likewise, the flue gas recirculation fan had deposits on the blades, but this had no operational impact, so that future cleaning can be limited to scheduled maintenance.

In general, the condition of the equipment at Pavisa after one year of operation is quite encouraging for the life of the system. Although the experience is invaluable for the design and construction of larger TCR systems, no fundamental changes to the approach demonstrated at

Pavisa will be necessary in the future. The only equipment that had to be improved after the startup of the TCR system were the refractory damper that closes the oxy-fuel flue gas port during TCR operation and the compressed air supply of the system. A number of refractory samples from various vendors were removed during the inspection shutdown that are currently being analyzed.

Figure 1: Picture of Regenerator at the beginning of the Inspection

GLASS QUALITY

The most important performance measure for any combustion system on a glass furnace is the ability to produce the desired glass quality at the requested pull rate. As stated above, Pavisa uses Furnace 13 to produce a broad palette of containers and this means demonstrating production with many different glass colors and types, some clear with a low seed count and some with a "rustic" appearance with visible bubbles.

During the design stage of the OPTIMELT system, many CFD models were completed at Praxair R&D to learn how to influence the furnace temperature profile with the syngas combustion system. Although a CFD model is always a simplification of a more complex reality, this effort confirmed that the TCR flames can be adequately shaped and it provided a basis for the startup and optimization team how to adjust the heat transfer to the needs of the production. As a result, the startup of the system was quite successful and no production was lost. To understand the changes between oxy-fuel firing and TCR operation a number of tests were run for different glasses. The furnace was first deliberately operated for a few days with oxy-fuel combustion using the existing six Praxair Wide Flame Burners in the side walls while operational and quality data were collected as part of the normal Pavisa production and quality control. Then the operation was switched to OPTIMELT with the same glass type and pull rate so that the collected data could be easily compared to the oxy-fuel operation. In general, the results from many test runs indicated that TCR fuel and oxygen savings were between 15 and 18%, and glass quality was well within the required specifications.

Most container glass companies do not vary their production as often as Pavisa and performance on flint glass for clear food and beverage containers is of interest. To show the performance under these conditions a test with clear flint glass was run in August 2015 on Furnace 13 and the results are presented in the subsequent paragraphs. The production run on fine flint glass was started with oxy-fuel and stable conditions were established in the initial four days. For the next four days, samples and quality data were evaluated at steady state conditions. Then, the furnace was switched to TCR operation and conditions stabilized before the TCR samples and data were evaluated at steady state conditions for another four days. No change in batch formula was made (or necessary) during this evaluation period and care was taken to keep operational parameters the same.

Key operational data is summarized in Table 1 below. Consistent with earlier experience, it was possible to increase the production during the operation of TCR, in this case from 50.5 tpd on oxy-fuel to 52.5 tpd pull rate (+4%). The total energy consumption of furnace and forehearth was measured to include any effects of a glass temperature variation, but to compare the total energy consumption the TCR data were corrected to the lower pull rate.

Table 1: Operating Data for Quality Comparison Tests

Parameter	Unit	Oxy-fuel Average	OPTIMELT Average	Variation
Pull Rate	t/d	50.5	52.5	4%
Cullet Rate	% dry	36	36	-
Excess Oxygen	% wet	2.7	2.3	-0.4%
Furnace Hot Spot Temperature	°C	1529	1524	-5°C
Melter Bottom Temperature	°C	1314	1312	-2°C
Total NG Used (*corrected to 50.5t/d pull)	$m^3{}_N/h$	375	308*	-18%

As to be expected with a cycling system that switches between two regenerators, TCR shows slightly larger changes of furnace pressure and glass level in comparison to oxy-fuel operation. However, similar to air regenerative systems this does not affect production and a comparison shows that the TCR systems pressure swings are less than those of air furnaces. The glass quality was found to be comparable between TCR and oxy-fuel operation. Specific data is presented in Table 2.

Table 2: Results of Quality Comparison Tests

Parameter	Unit	Oxy-fuel Average	OPTIMELT Average	Variation
Seed Count	1/ounce	23	33	10
Bottles with Stones	%	1	1	-
Dominant Wavelength	nm	571.6	568.9	-2.7nm
Transmittance	%	81.00	80.97	-0.03%
Fe^{2+}/Fe^{3+} redox ratio	-	0.278	0.285	+0.07
Fraction of Fe_2O_3	%	78.25	77.80	-0.45%

The seed count increased during operation with the TCR system, but stayed well within the specification of 60 counts/ounce for the category of seeds and larger bubbles (blisters). A slight increase in bottom glass temperatures would have lowered the count to the level of the oxy-fuel operation and later data showed seeds as low as 10 counts/ounce at a bottom temperature of 1321°C when operating the TCR system. All other quality data showed only slight changes.

Of specific interest is the redox ratio for the glass, which could point to problems with creating a furnace atmosphere that is reducing. During the design of the syngas burner, this issue was specifically addressed with extensive CFD modeling. The operating data indicate that this work was successful and that there is no indication of the OPTIMELT syngas burners creating large areas of reducing atmosphere over the glass melt that would negatively influence the redox ratio.

NOx AND CO EMISSIONS

The Praxair and Pavisa technical team jointly conducted flue gas NOx and CO emissions measurement under continuous operations in early June 2015 during a one-week period. Gas samples were taken in the RFG piping loop and before the main stack, thus air streams used for flue gas cooling or furnace pressure control in the main stack did not affect the emissions measurement. Dry gaseous concentrations (CO, CO_2, O_2, and NOx) were measured by a well-calibrated Sick-Maihak S710 gas analyzer. During that measurement period, the furnace was mostly pulled at 40 tpd with a cullet ratio of 30%.

To broadly test the TCR system emissions characteristics, various furnace and syngas burner operating parameters were varied systematically. These parameters included furnace control pressure, excess oxygen in the flue gas, flow rate of the RFG, and the flow rates of oxygen used by the syngas burner. Other burner parameters for both the fuel and oxygen were kept the same. Changes in the value of the test variables were intentionally set to be very wide for test purposes, for example, flue gas oxygen concentrations were altered between about 2% to 12% (dry).

Total flue gas flows were calculated using a validated furnace mass balance model. The model's calculated O_2 and CO_2 concentrations at known pull rate and fuel flow were forced to match with the measured O_2 and CO_2 concentrations, by changing the flows of the unknown furnace air leaks iteratively. The mass or volume based NOx and CO emissions data could then be calculated based on the model's outputs in total flue gas flow, furnace nitrogen content, and flue gas water vapor concentrations.

Concurrent to the Praxair emissions test, a certified local contractor was hired to perform several emissions measurement at the main stack according to EPA 7E Method. The emissions numbers measured by the contractor were in general lower than those determined from the measurements by Praxair/Pavisa at the RFG pipe location and using the calculation method as described in the sections above. In the following, only NOx and CO data measured by the Praxair/Pavisa team are presented.

The measured NOx emissions are plotted against the calculated nitrogen concentration in the wet flue gas under normal TCR operating conditions in Figure 2. Each data point represents a time average of measured NOx during a measurement period between 5 to 25 minutes when the TCR system was either left or right firing. The NOx variation shown were the result of left and right firing differences, the tuning of flue gas excess oxygen, changes in furnace control pressure, and the distribution and adjustment of oxygen flows. As can be seen, the furnace nitrogen concentrations were between 20 to 31 % wet due to a large ambient air infiltration into this small furnace and the measured NOx ranged from 0.3 to 0.75 kg/t (metric ton). Reducing the furnace control pressure increased the air ingress into the furnace which resulted in a high furnace nitrogen content of about 31%.

Figure 3 plots the measured CO concentrations corrected to 8% reference oxygen against the measured excess oxygen percentage in the flue gases when the TCR system was either left or right firing. It can be seen that CO was below 80 ppm when the flue gas excess oxygen was above 4% (dry), and lower excess oxygen operation increased the peak CO concentrations up to about 130 ppm. The shape of the CO-O_2 curve is consistent with typical oxy fuel combustion where approximately 2% wet O_2 (about 4% dry) in the flue gas is normally recommended. When converted to a corresponding weight based unit, measured CO emissions ranged between 0.04 to 0.2 lb/ston (short ton) or 0.02 to 0.1 kg/t of glass pulled.

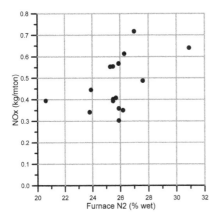 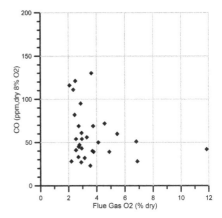

Figure 2: Measured NOx vs. Furnace Nitrogen   Figure 3: Measured CO vs. Excess Oxygen

The NOx data (kg/t) measured at Pavisa must be scaled for the effects of nitrogen concentration in the furnace, ambient pressure, and furnace size/pull in order to compare with NOx data obtained from other oxy fuel fired furnaces. NOx emission from oxy-fuel fired furnaces is known to increase approximately linearly with the nitrogen concentration in the furnace both from the NOx formation theory and from actual data. Due to high elevation, atmospheric pressure at Pavisa is only about 0.76 atmospheres. At a same furnace nitrogen percentage, the partial pressure of nitrogen in the Pavisa furnace was lower when compared to that of a furnace located near sea level where atmospheric pressure is 1 atm. For example, a data point at 25% nitrogen at Pavisa would correspond to 25 x 0.76 = 19% nitrogen at sea level.

The other factor one needs to be aware of is the effect of furnace size on NOx emission. Small furnaces have higher specific wall losses and hence higher specific fuel consumption. For example, fuel consumption in the 50 tpd Pavisa furnace under oxy-fuel firing was about 6 GJ/t as compared with a typical 4 GJ/t range for an oxy-fuel fired 300-400 tpd container furnace. Thus, specific NOx emission (kg/t) from small furnaces tends to increase due to their higher specific fuel consumption.

In Figure 4, NOx emission from the Pavisa furnace under TCR operation is compared with the data of Praxair's Low-NOx JL and A-burners, previously collected from other oxy-fuel fired container glass furnaces. The comparison is based on the partial pressure of nitrogen in the furnace. It indicates NOx from TCR operation was in the range of measured values of Praxair's Low-NOx JL oxy fuel burners, even under the indicated high furnace nitrogen partial pressures

(0.16 to 0.24 atm). Optimization of the TCR burner is expected to reduce NOx emissions even further.

Comprehensive reviews of glass furnaces air emissions and compliance issues had been published recently [5]. According to that review, one of the most stringent glass emissions regulations in the US is in San Joaquin Valley, CA. Based on SJVUAPCD Rule 4354 [6], for oxy fuel fired container glass furnaces NOx limit is 1.5 lb/ston (or 0.75 kg/t) and CO limit is 1.0 lb/ston (or 0.5 kg/t). In the EU, one of the agencies regulating NOx emissions is the European IPPC Bureau. NOx emissions for container glass furnaces can be found in reference [7] as BAT, which shows for oxy fuel container furnaces the NOx limit is between 0.5 to 0.8 kg/t (1.0 to 1.6 lb/ston) depending on furnace nitrogen concentrations.

Our data analysis for the measured NOx data indicates that a NOx level at or less than 0.6 kg/t (1.2 lb/ston) can be achieved under TCR system operating conditions, assuming no nitrates are used in the batch formulation. It is our expectation that by working closely with glass plant operators, the technology can meet future emissions requirements imposed by various agencies while recovering flue gas energy from oxy fuel furnaces.

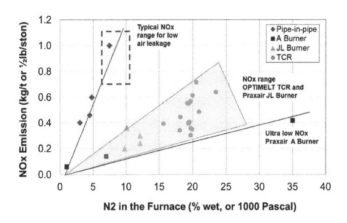

Figure 4: Comparison of NOx emissions from OPTIMELT to oxy-fuel burners

ENERGY SAVINGS ON LARGER FURNACES

The results were used to validate the TCR performance model that allows predicting the expected energy savings for larger furnaces operating under different operating conditions. For example, low cullet ratio and high glass melting temperature tend to increase the system's fuel savings because the flue gas has more energy to be recovered due to its larger volume and higher temperature. Bigger furnaces have lower wall losses that increase the recoverable energy in the flue gas relative to the total heat input.

Figure 5 summarizes various recovery performance analyses for commercial glass furnaces. These were either existing oxy-fuel furnaces or air-fuel furnaces that are scheduled for rebuilt. For a typical large container furnace in the range 300 to 400 tpd pull, fuel savings with the TCR system are projected to be between 18 to 22 % depending on baseline oxy-fuel furnace operations and 26 to 31 % compared to air regenerative furnaces. The savings relative to recuperative furnaces would be higher. One interesting aspect of the system implementation is the increase of furnace pull in combination with a total rebuilt when the site space is limited or the capacity of the oxygen supply system insufficient for the new larger furnace. The small footprint of TCR technology offers opportunities to integrate a much larger furnace into an existing building while reducing operating costs with oxy-fuel technology and advanced heat recovery.

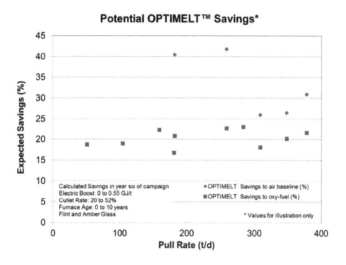

Figure 5: Predicted Energy Savings versus Furnace Pull Rate

CONCLUSIONS

The thermochemical regenerative heat recovery process has been demonstrated in a 50 tpd commercial glass furnace at Pavisa since September, 2014. The process has been shown to be able to operate reliably without any production losses or glass quality issues under frequent product changeover schedules. The technology is now in the early commercialization stage and the tools and engineering for scale-up are being developed. As a direct result of the projected fuel savings of 26 to 34% compared to state of the art air fired regenerative furnaces, the CO_2 emissions are reduced by the same percentage. Recent emission measurements indicated that NOx emissions of the process are less than half of the conventional oxy-fuel burners and that it can meet stringent future NOx and CO emissions regulations. There is potential to reduce emissions further through overall furnace optimization including fine tuning of the syngas burners.

REFERENCES

1. Kobayashi, H., Wu, KT, Bell, R.L. (2014), "Thermochemical Regenerator: A High Efficiency Heat Recovery System for Oxy-Fired Glass Furnaces", DGG/AcerS Conference, Aachen, 28 May 2014.
2. Gonzalez A., and Solorzano, E., et. al. (2014), "Commercial Demonstration of ™ Thermochemical Heat Recovery for Oxy-Fuel Glass Furnaces", 29th A.T.I.V. Conference and 12th European Society of Glass Conference, 21-24 September 2014, Parma, Italy
3. Gonzalez A., and Solorzano, E., et. al. (2014), "Optimelt Regenerative Thermo-Chemical Heat Recovery for Oxy-Fuel Glass Furnaces", 75th Conference on Glass problems, 3-6 November 2014, pp. 113-120.
4. Iyoha, U., et. al. (2015), "Thermochemical Regenerator System Proves Itself at Pavisa", Glass International, May, pp. 29-30.
5. Ross, C.P. (2009), "Current Glass Furnace Air Emission Compliance Issues", Ceramic Engineering and Science Proceedings, Vol. 30, Issue 1, pp. 215-226.
6. SJVUAPCD (2011), "Glass Melting Furnaces", Rule 4354, Amended, 19 May 2011, pp. 4354-5 to 4354-6.
7. ISSN 1977-0677 (2012), "Official Journal of the European Union", L 70, English Edition, Volume 55, 8 March 2012, p. 19.

Batching

OPTIMIZATION PROGRAM FOR BATCH WETTING OFFERS BENEFITS IN FURNACE LIFE, ENERGY EFFICIENCY, AND OPERATIONAL RELIABILITY

Thomas S. Hughes, F. Philip Yu, and Blaine Krause
Nalco, An Ecolab Company
Naperville, IL USA

ABSTRACT

This present paper highlights the results of applications of Dry Batch Optimizer (DBO) technology in the container glass industry. The authors will discuss proper use of this technology to aid in the effectiveness of batch wetting, allowing glass manufacturers to take full advantage of the benefits of batch wetting while minimizing or eliminating its drawbacks. Pre-wetting of batch material prior to its addition to the furnace is a common practice, and most glass manufacturers agree on its benefits: reduced dusting inside and outside the furnace, extension of furnace life, reduced fouling of regenerators, and better control of particulate emissions from the furnace. Nalco has now documented information showing that we can achieve additional benefits such as: improved homogeneity of the batch mixture, reduction of excess moisture and associated energy benefits, improved batch pile shape and resulting melting efficiency leading to additional energy savings, improved cleanliness of mixers with reduced torque / wear and tear, and finally, improved safety resulting from reduced dusting as well as reduced need for operator intervention to clean the batch feed system.

BACKGROUND

Glass manufacturers use a mixture of dry materials, such as silica sand, soda ash, sulfate, limestone and dolomite as raw materials for batch processing. These raw materials are mixed in the appropriate ratios, and transported into the furnace for melting, forming a uniform, liquid mixture prior to exiting the furnace for forming into the final product. In most cases, cullet (recycled glass) is added to the batch mixture, at various ratios, prior to charging the furnace. Batch recipes vary, based on color and quality requirements of the end-use for the glass.

Most glass manufacturers add water, via spray-nozzles at a wet screw, or in a large mixer, to achieve a target moisture level of the material as it enters the furnace at the charger. The addition of water is designed primarily to reduce dusting, both inside and outside the furnace. Suppression of dusting is important for several reasons:

- To minimize dusting as an inhalation hazard for workers in the area.
- To minimize particulate emissions for environmental reasons.
- To minimize dusting inside the furnace (often referred to as "batch carryover") – harmful for the following reasons:
  - Particulate matter from the batch can accumulate in the crown section of the furnace, reacting with and damaging the refractory, potentially adding to maintenance costs and/or shortening furnace campaign life.
  - As this material breaks loose or melts down into the molten glass, it can create irregularities, leading to glass quality problems.
  - Particulate matter in the gas stream may accumulate in the exhaust flues and/or regeneration checkers, creating back-pressure and disrupting normal gas flow through the furnace and exhaust system.

Batch wetting, in the proper proportions, can be used to control dusting and carryover, helping to mitigate the challenges listed above. While the benefits of batch wetting are widely recognized, there are also potential drawbacks:

- Evaporating the added water represents a potential additional energy cost. Though some addition of water may have a neutral energy effect due to the exothermic nature of the hydration of soda ash, excess water addition may represent an extra energy load.
- Batch wetting sometimes results in "caking" or "adhesion" of the batch material to the chutes and hoppers in the batch feed system, potentially resulting in plugging of mixers, wet-screws, chargers, and other parts of the batch transport and feed systems. This adhesion can sometimes interrupt normal operation of the batch feed system, with the result that furnace operations personnel are diverted from proactive management of the furnace to more rudimentary cleaning activities related to the batch feed system.

We have developed a program using a process additive which enhances the wetting and batch mixing process, and minimizes the potential disadvantages of batch wetting. The program enhances the wetting process, accelerating and improving the distribution of moisture through the batch material. The improvement in moisture distribution and mixing provides several benefits for the batch handling process:

- Reduced dusting and batch carryover.
- Prevention of caking and adhesion of batch material to feed system components, such as chutes, hoppers, wet-screws, mixer blades, chargers.
- Improved homogeneity of batch mixture, contributing to improved eutectic melting and melting efficiency.
- Improved texture of batch and shape of batch piles / logs, positively influencing angle of repose in the furnace and the associated benefits to melting efficiency.

EARLY DEVELOPMENT PROCESS FOR DRY BATCH OPTIMIZER PROGRAM:

The DBO program was originally developed in response to a client's request. A North American container glass plant was using the cooling water treatment services and was also adding a surfactant developed by another supplier as a batch wetting aid. Due to microbiological fouling of the surfactant feed system (see Figures 1 and 2), program results were unreliable, and batch wetting nozzles were frequently plugged, requiring unscheduled maintenance of the batch wetting system and affecting the plant's ability to achieve their moisture targets. Though adding a biocidal agent to the mix tank was considered, this would have resulted in additional costs, and it was unclear what effect this would have on the batch wetting process.

Figure 1: Mix Tank used for mixing & feeding previous wetting agent

Figure 2: Wetting water foulants plugged nozzles at the wet-screws

PHYSICAL APPEARANCE

Physical State	Solids	Quantity of Solids	
Turbid liquid	Flocs	Heavy	

Analyte	Result		Test Method
AEROBIC BACTERIA			CB22010
Total Viable Count @ 35°C	320000000 est.	CFU/mL	
Pigmented Bacteria	1 Type		
Mucoid Bacteria	Not Detected		
Total Coliforms	360000	CFU/mL	
E. coli	<100	CFU/mL	
Pseudomonas spp @ 35°C	380000 est.	CFU/mL	
ANAEROBIC BACTERIA			CB22016, CB22018
Sulfate Reducing bacteria	>100	CFU/mL	
FUNGI			CB22015
Mold	13000	CFU/mL	
Yeast	<10	CFU/mL	

Figure 3: Differential microbiological analysis of suspected bio-foulants found in mix tank

As an alternate approach, we conducted a screening of many different formulations of wetting agents and identified some products, which would improve batch wetting, without contributing to microbiology growth or fouling. Early in the development process, the problem of bio-fouling was eliminated, and feed system reliability was restored. (See Figure 4.)

Before formulation change: Slimy deposits growing in mix-tank

After formulation change: No bio-fouling and water is clear to bottom of tank

Figure 4: Condition of mix-tank before and after wetting agent formulation change

Additional benefits of the formulation change:
- In the course of identifying alternative surfactant formulations, only formulations with negligible sulfur content were considered. In this way, sulfur contribution was eliminated, thus reducing any contribution to SO_x emissions from the furnace.
- The formulation change improved mixing and reduced caking and adhesion of the batch in the feed systems, resulting in less need for cleaning of the batch feed system by operations personnel.

Figure 5: Improved cleanliness in the screw feeder of the after treating with new formulation

Follow-up R&D Work to Find the Optimal Solution:

Following the initial success in resolving microbiological fouling issues and improving upon some of the performance characteristics of the wetting agent, it was important to ensure that the most cost-effective formulation was chosen, one that would maximize batch wetting benefits. The first step was to compare a series of surfactant formulations with regard to their impact on surface tension.

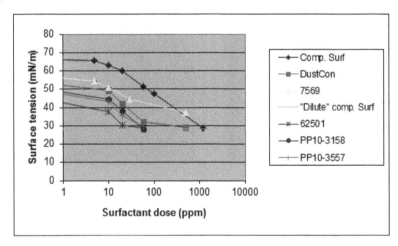

Figure 6: Examples of testing of various surfactant formulations for surface tension impact

Second, a laboratory simulation of the batch mixing and wetting process was needed to evaluate different formations. A laboratory simulation (Figure 7) of the wet-screw feeder was set up to conduct quantitative measurement of the torque generated during the dry batch mixing process, along with the wetting agents' impact on "flowability." A comprehensive screening was also initiated to evaluate other alternate chemicals to identify the product that generated the most optimal performance. A Precision Stirrer with torque controller and measurement was employed to drive an impeller assembly to simulate the screw feeder motion in the glass plant. The torque output was transmitted through an analog to digital signal converter to a data logger.

Figure 7: Laboratory setup (side and top views) to simulate dry batch screw feeder.

Batch mixture was maintained at 3-5% moisture level, and mixed with various wetting agents to determine the product with the best impact on torque reduction. Product A was the first program applied to resolve the plugging problem at the packaging glass plant. Product B was later identified to provide improved performance over the previous chemistry. Ultimately, a third formulation, Product C, provided the overall best performance, and was also observed to be the best option related to ease of feeding and mixing. With 3% moisture content in the dry batch mixture, Product C generated the lowest torque, which would contribute to a smoother mixer operation and lower energy consumption (Figure 3). Product C was named the Dry Batch Optimizer (DBO), and the plan was to proceed with additional in-plant evaluations.

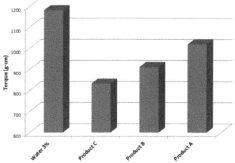

Figure 8: Torque measurement of dry batch mixing in a laboratory scale stirrer with double impellers to simulate screw feeder. All combinations were conducted under 3% moisture

Operational Impact of DBO in a Plant with No Previous Use of Batch Wetting Surfactants:

Following this initial development work in container glass plants, where batch wetting surfactants were previously in use, a test site was needed to help us determine the impact of using DBO where no wetting agent was previously in use. We agreed to conduct an evaluation with another container glass plant to evaluate the operational impact of the DBO program, and agreed on Key Performance Indicators we would track:

- Moisture levels
- Pull-rate / cullet ratio, etc.
- Cleanliness of wet-screws, hoppers, chargers
- Batch pile shape and distribution within the furnace
- Opacity readings from stack

In general, the program was well-received, with improvements noted in cleanliness of the batch feed system, batch pile formation, and general consistency of batch feed and level control in the furnace. No adverse impact was observed on quality or on emissions. In order to get clear confirmation of improvements in operations, we conducted some "off-tests," where feed of the DBO program was discontinued, and the monitoring of operational characteristics continued. In multiple cases, system cleanliness and batch pile shape suffered each time DBO feed was off.

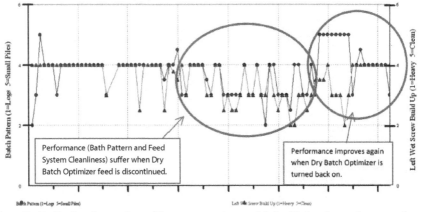

Figure 9: Qualitative observations of furnace operation at trial site #3 showing performance drop when Nalco DBO program was turned off. (Blue line/dot ~ Batch Pattern, Green line/dot for left Wet Screw buildup; 1 = bad and 5 = good)

Over the next few months, the dosage was optimized and operators became more comfortable with the performance of DBO program. As good performance continued, operators were able to steadily reduce the moisture content in the dry batch operation while maintaining feed system performance and batch consistency in the furnace. The average batch moisture was reduced safely from 3.9% to 2.9% over a six month period as operators continued to fine-tune the system. DBO made this possible by improving the way moisture distributed through the batch mixture, resulting in more even mixing (Figure 10).

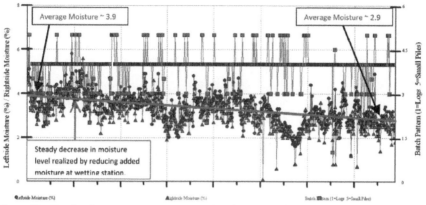

Figure 10: Dry batch moisture content was reduced by 25% from optimizing the Nalco DBO program in the furnace of trial site #3. (Blue: Left side Moisture; Green: Right side Moisture; Purple: Batch Pattern; 1 = bad and 5 = good)

Buildup of caked-on batch material in the feed chutes and hoppers often leads to flow restrictions and level control problems in a furnace. Prior to the implementation of Nalco DBO program, operators estimated that an average of 20 minutes per shift was spent on cleaning out the feed system to prevent problems. During the seven month evaluation period, the batch chutes, augers and hoppers for the chargers of the furnace stayed considerably cleaner without significant intervention. Figure 11 shows clean batch feed system during an inspection.

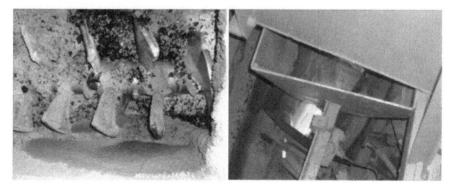

Figure 11: Clean screw feeder and charger chute observed during inspection at trail site.

During the evaluation, the batch pile shape and formation in the furnace were reasonably good with just water as wetting agent (Figure 12a). However, the operators were challenged with adjustments to prevent "logging" and keeping good, distinct batch piles. With incremental adjustments in dosage, the evaluation team found an optimal dosage of Nalco DBO, which has maintained good batch pattern (Figure 12b), but with less intervention needed by the operators.

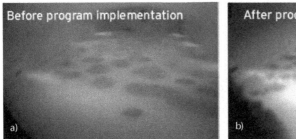

Figure 12: Batch pile shape / formation before and after Nalco DBO program at trial site.

GAINS IN MELTING EFFICIENCY DEMONSTRATED ATFLOAT GLASS PLANT

Although the operational improvements offered by Dry Batch Optimizer were well-received, it was still important to identify the program's direct impact on melting efficiency and energy use. In some cases, container plants had the challenge of frequent job changes, etc., and did not offer the steady-state operation which would facilitate direct measurement of the energy impact of the program. Nalco began work with a North America float glass plant to structure an evaluation which would include tracking the impact on energy use and melting efficiency.

This plant had experienced challenges with internal dusting and batch carryover in one of their furnaces, along with high defect rates when carryover was occurring. For this reason, prior to our evaluation, they were operating at over 6% moisture in their batch in order to suppress dusting. Operations personnel were also experiencing challenges with flow-restrictions and plugging of the batch feed systems, requiring frequent cleaning and maintenance on the chargers, hoppers, and associated equipment. The process was so labor intensive that the plant had installed a PTFE coating for some of their chargers in an attempt to keep batch material flowing more smoothly.

In developing the guidelines for the trial, several Key Performance Indicators (KPI's) were identified, and two months of baseline data were collected:

- Water addition
- Batch moisture
- Thermal profile of melters
- Stack output
- Fuel usage
- Defect counts

The batching process included a mixer, followed by a conveyor system, which transferred wetted, mixed batch to a silo with approximately eight hours or holding time. Dry Batch Optimizer was blended with the batch during the water addition stage of the mixing process. During baseline data collection, moisture at the chargers was typically over 6%. Early in the evaluation, as the DBO program began to help improve overall batch handling and flow through the system, furnace operations personnel began to reduce batch moisture (Figure 13). Several step-wise moisture reductions were made, until the system began to show increased defects, below 3% moisture. Moisture was subsequently increased, and the long-term moisture level target was moved to 4%.

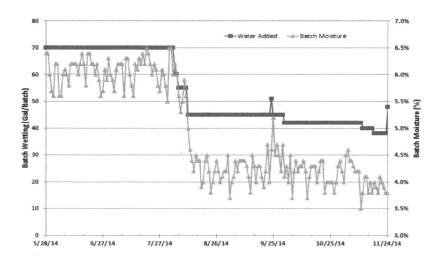

Figure 13: Stepwise decreases in added moisture to the furnace with no adverse effect on batch carryover or product quality.

As the trial progressed, a significant reduction in charger buildup was observed. As a result, operators spent less time and effort on weekly scheduled preventative maintenance (PM) to clean the chargers. At one point during the trial period, a problem with the conveyor belt occurred, and the plant was forced to hold wetted batch in the mixer for almost 6 hours before operation resumed. In the past, such an interruption in operation would have caused the mixture to clump inside the mixer and plug the outlet. Surprisingly, when the conveyer belt resumed operation, the wetted batch mixture that was held for this extended period of time came out of the mixer without clumping or plugging. A similar event in the past would have required several hours of manual cleaning of the mixer by a maintenance crew, removing clumps of hardened material in order to resume production.

Increased furnace temperature stability

As the DBO program fully integrated into furnace operation, a more homogenous flow in batch mixture resulted in improved stabilization in furnace bottom temperature. Previously, there were fluctuations in the temperature profile of the molten glass as the dry batch entered the furnace, which meant additional charger adjustment was required. The DBO program improved the batch wetting process, creating a homogeneous flow in batch charging with stable furnace bottom temperature and no additional fuel consumption. As a result, there was no need for the operators to adjust charger gate settings to compensate for uneven heating.

The three bottom sensors in the Melter temperature profile showed a marked improvement with the DBO program. The crown temperature showed a downward trend, which implied an impact on melting efficiency inside the furnace (see Figure 14). The stack output showed no significant change under the Nalco DBO program.

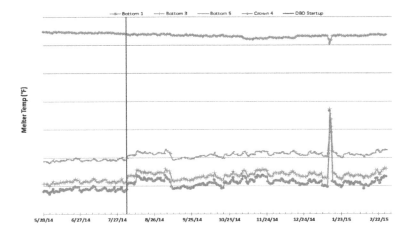

Figure 14: Improved stability in furnace temperature profile, positive impact on melting efficiency.

Fuel Savings

During the trial period, there was a downward trend in overall plant fuel usage. Figure 3 shows that fuel usage (in MMBTU/ton) was reduced by 3.4% on average when running the DBO program under stable conditions. An average 0.21 MMBTU/ton of savings was attributed to the Nalco DBO program implementation (See Figure 15).

It was noted that the overall measured impact on energy was considerably more than the energy savings attributable to reduced moisture levels and the energy savings associated with vaporizing that moisture (estimated at 0.055MM BTU/Ton). The additional energy savings are likely attributable to other factors impacting melting efficiency, such as:

- More effective mixing, leading to batch homogeneity.
- More complete hydration of the soda ash, an exothermic reaction which heats the batch internally.
- Modified texture of the batch, leading to increased angle of repose and its impact on surface area of the batch logs in the furnace, as well as improved ablation.

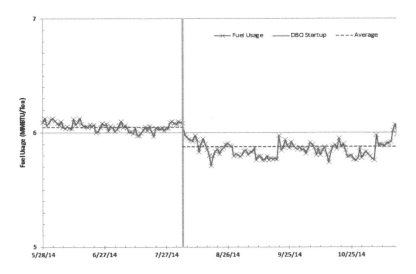

Figure 15: Decrease in fuel used per unit of production as a result of the DBO program.

Glass Quality Maintained During Evaluation

Nalco DBO did not have any adverse impact on the glass production quality. Figure 16 indicates a stable trend in quality in quality defects. The occasional spikes recorded in the trend chart were caused by operational changes that had no correlation with the DBO program. Although there were signs of slight gain in glass quality, the batch wetting additive had no overall impact on edgelites, optical distortion and glass chemistry. No significant stone lines or carryover issues were observed during the deployment of DBO program. When the batch moisture was reduced below optimum to 3.8%, an increase in defects was detected. As water was added back to batch wetting to 4% moisture, the defects dropped below 2 per 100 ft².

Impact on Key Performance Indicators are:
- 37% reduction in average moisture content of the dry batch mixture (Figure 13).
- Improvement in melter thermal profile clearly demonstrated (Figure 14).
- Fuel savings on average of 3.4% or 0.21 MMBTU/ton (Figure 15).
- Glass quality maintained; no quality control issues detected during the Nalco DBO program based on defects per 100 ft2 (Figure 16) .
- Reduced routine maintenance; chargers remained relatively clean without significant deposition. Only slight pluggage occurred on the metal grating on top of the charge hopper.
- No adverse impact in stack output / emissions detected after the startup of batch wetting program.

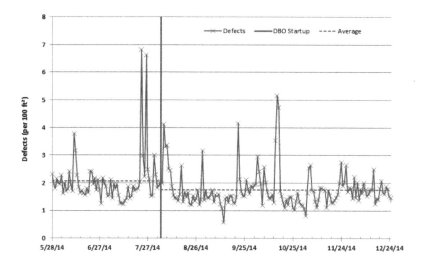

Figure 16: Total defect counts during evaluation period.

ECONOMIC RESULTS

The Nalco DBO program is designed to improve the batch wetting process to yield more homogeneity of the batch mixture, while reducing water usage needed to suppress dusting. Based on improvements realized within the evaluation period, the detailed calculations of total cost of operation reduction are listed in Table 1. The most important finding of the evaluation was related to melting efficiency. When discussing batch wetting, there is general agreement in the industry that reducing excess moisture will yield energy savings. These energy savings are simply related to the latent heat of evaporation and can be easily calculated using a steam table.

However, this evaluation showed a significant impact on melting efficiency and energy usage beyond what could be expected from moisture elimination. About 70% of the energy savings realized was not attributable to evaporation of the excess moisture. Rather, they are related to other factors in overall furnace melting efficiency. These other factors include:

- Homogeneity of the batch mixture, and the associated encouragement of eutectic melting.
- Optimization of moisture distribution through the batch mixture, potentially influencing soda ash hydration effectiveness and improving conductance of heat throughout the batch material.
- Influence of DBO on the texture of the batch mixture, allowing increased angle of repose for batch logs and contributing to consistent furnace circulation patterns.

Glass Furnace Production (tons/day)	718
Operator hours saved per week (cleaning feed system/uplugging chutes, etc.)	1.67
Reduced labor associated with system component cleaning/year @ $13/hr.	$ 1,129
Reduced maintenance cost for mixer clean-up (not yet quantified)	TBD
Reduced maintenance cost for charger rebuilds (not yet quantified)	TBD
Reduced Labor and Maintenance Subtotal	$1,129 +
Batch Moisture prior to Nalco DBO program - %	6.5%
Batch Moisture with Nalco DBO Program - %	4.0%
Water Reduction with Nalco Program lbs/day	35,900
Cost of Water - $/Kgal	$ 2.50
Annual Water Savings with Nalco DBO Program	**$3,926**
Fuel Reduction related to reduced moisture with Nalco DBO Program-BTU/day	39,920,800
Cost of Fuel - $/MBTU	5.16
Annual Fuel Savings due to reduced moisture with Nalco Program-$/year	**$75,187**
Actual measured fuel energy use reduction (0.21 MM BTU/Ton) -- MM BTU's	55,035
Total Fuel Savings Measured (per year)	**$283,979**
Calculated Fuel Savings due to factors other than moisture reduction	**$208,792**
Total Annual Fuel and Water Savings with DBO program	**$287,905**
Total Annual Savings with Nalco Program - $/year	**$289,034**
Typical program cost / year	**$75,000**
Net TCO Reduction Savings with Nalco DBO Program	**$214,034**

Table 1: Economic impact of DBO Implementation

REFERENCES

1. Davis, D. H. and Holy, C. J. To wet or not to wet – that is the question – part a, Ceramic Engineering and Science Proceedings (2011), 32.
2. Verheijen, O.S. Thermal and chemical behavior of glass forming batches, 2003 Dissertation of Technische Universiteit Einhoven, Netherland
3. Pecoraro, G. Method of making float glass. International Patent ZA7108362, 1973

EFFECT OF BORATE RAW MATERIAL CHOICES ON THE BATCH REACTIONS OF
ALKALI-LEAN BOROSILICATE GLASSES

Mathieu Hubert, Anne Jans Faber
CelSian Glass & Solar
Eindhoven, the Netherlands

Simon Cook, David Lever
Rio Tinto
London, UK

ABSTRACT
     Borosilicate glasses are of utmost importance in a large number of commercial
applications, and are among the most largely produced types of glass. The boron, integrated
within the structure of these materials, confers outstanding properties to the glass and is
irreplaceable for numerous glass compositions (such as low thermal expansion lab-ware or
display glasses).
     For industrially produced borosilicate glasses, different types of borate raw materials can
be chosen by the manufacturers (e.g. borax, boric acid, colemanite). The selection of the type of
borate employed for the glass batch depends notably on the type of borosilicate and the exact
composition of the raw material itself (e.g. the sodium-containing borax would not be used for
alkali-free glasses) as well as on economic considerations. The choice of the borate raw materials
may also have an impact on the melting process itself, e.g. on the melting kinetics and the
evaporation processes. In this study, the influence of the choice of the borate raw materials upon
the melting behavior of alkali-lean borosilicate glasses is presented.
     Industrial alkali-lean borosilicate glass batches were melted using High Temperature
Melting Observation System and Evolved Gas Analysis setup (HTMOS/EGA). It is shown that
the type of borate raw material employed has an effect on the melting and fining behavior of the
glass. Notably, the influence of hydrated vs. anhydrous borate raw material (boric acid vs. boron
oxide) is highlighted.

INTRODUCTION
     Boron is an essential component in a variety of industrially important families of glasses,
such as laboratory ware, cookware, insulation glass-wool, (reinforcement) textile fiberglass (e.g.
E-glass, S-glass), display glass… Boron is added as it confers unique properties to these glasses,
such as improved thermal stability, chemical durability, or reduction of electrical conductivity. In
the glass melting process, addition of borates leads to reduction of melting temperatures and of
the melt viscosity at high temperatures, which is a significant advantage in terms of energy
consumption [1, 2].
     Though the role of boron in the final borosilicate glass (i.e. on the final product) is
independent of the borate raw material used in the batch, different types of borate raw materials
may be used. The choice of the borate raw material depends on several factors. First, the final
composition and desired properties of the glass may limit the use of some of these raw materials.
For instance, sodium borates cannot be used to produce borosilicate glasses with low electrical
conductivity, i.e. alkali-lean or alkali-free glasses such as E-glass or some display glasses. Other
important criteria considered for the selection of the borate raw material include purity (e.g.
mineral vs. refined borate) and costs.
     The main types of borate raw materials used in the glass industry for production of
borosilicate glasses include notably:

- Sodium borates:
  - Borax decahydrate: $Na_2O.2B_2O_3.10H_2O$
  - Borax pentahydrate: $Na_2O.2B_2O_3.5H_2O$
  - Anhydrous borax: $Na_2O.2B_2O_3$
- Boric acid: H_3BO_3, or $B(OH)_3$
- Boric oxide: B_2O_3
- Mineral borates:
  - Ulexite: $Na_2O.2CaO.5B_2O_3.16H_2O$
  - Colemanite: $2CaO.3B_2O_3.5H_2O$

As can be seen from the list of borate raw materials given above, another criterion that can characterize them is whether they are hydrated or anhydrous borates. Indeed, sodium borates can be either found as anhydrous borax or hydrated (deca- or pentahydrate) borax. In the same manner, anhydrous boric oxide has its hydrated counterpart, boric acid. Dehydrated borates are produced by fusion process to remove water of the hydrated raw materials. Therefore, batches prepared using anhydrous borate raw materials will release less water than batches prepared with their hydrated counterpart [3, 4]. By consequence, they typically require lower energy for melting, increasing furnace productivity and melting energy costs. In addition, the reduced water release from the batch may lead to lower borate volatility and dust emissions.

In this paper, the effect of the choice of borate raw materials on melting-in behavior of 3 different low-alkali glasses will be investigated. A special emphasis is put on the influence of anhydrous vs hydrated borate raw materials on the melting behavior of these glasses. The investigations have been carried out using high temperature melting observation combined to thermodynamic calculations.

MELTING EXPERIMENTS

The melting behavior of borosilicate glasses prepared with different borate raw materials was investigated using High Temperature Melting Observation System, coupled to Evolved Gas Analysis (HTMOS-EGA). An illustration of the setup and its principle is given in Figure 1. The actual setup is shown in Figure 2.

The batch to be investigated (typically 100 – 200 g) is placed in a silica crucible, which is then introduced in an electrically-heated furnace. The top of the crucible is sealed with a water-cooled lid, and a constant carrier gas is led into the crucible in order to simulate the industrial furnace melting atmosphere. The composition of the atmosphere above the melt is controlled via mass flow controllers and can be tuned to simulate different types of atmospheres, notably various water contents in the atmosphere (e.g. about 20 % water vapor for air-fuel fired furnaces, approx. 55 % for oxy-fuel fired furnaces).

The gases released from the batch/glass mix are continuously carried (together with the carrier gas) via heated lines to a FTIR and oxygen analyzers, for the analysis of the flue gas composition (EGA). The volume fractions of the released gases, typically CO, CO_2 and SO_2, are measured as function of time and temperature.

The experimental set up is also equipped with a high resolution CCD camera, taking pictures of the batch/glass melt in the silica crucible at regular intervals. At the end of the experiment, the images are compiled into a movie, and can also be used for detailed observation of the reactions taking place (e.g. onset of melting phase formation, measurement of foam amount, determination of the fining onset temperature). The combination of the video observation and of the EGA allows for a better understanding of the melting-in, fining and foaming behavior of batches. By comparing the videos and EGA results for, for instance, two batches containing different borate raw materials to prepare the same final glass, the impact of these borate raw materials can be investigated.

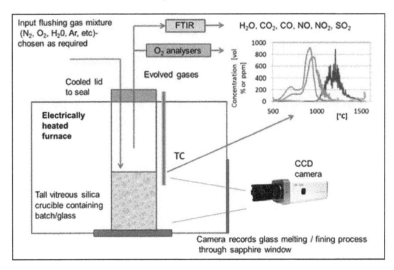

Figure 1. Schematic illustration of CelSian's HTMOS-EGA setup

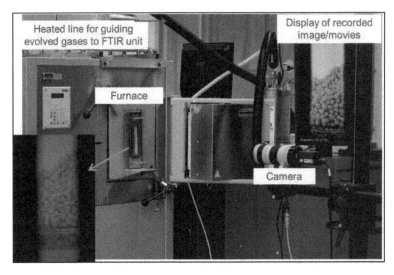

Figure 2. Picture of CelSian's HTMOS-EGA setup, with illustration of the video recorded during melting of a pelletized batch

In this study, three different borosilicate glasses have been investigated, respectively a low-dielectric (or low-κ) glass, an E-glass and a Duran®-type hard borosilicate. Indicative compositions for these 3 glasses are given in Table 1. These glasses contain different boron contents, with respectively 4.8 wt%, 10.4 wt% and 13 wt% B_2O_3 for E-glass, low- κ and Duran® glass, respectively. The E-glass and the low-κ glass contain less than 2 wt% alkali, while the Duran® glass contains 4 wt% alkali.

TABLE 1 Indicative compositions for the glasses investigated

Composition (wt%)	E-glass	Low-κ glass	Duran® glass
SiO_2	55 - 57	60.2	81
$Na_2O + K_2O + Li_2O$	0.6 - 0.8	1.5	4
B_2O_3	4.8	10.7	13
CaO	22 – 24	3	-
MgO	1.5 - 2.5	11.3	-
Al_2O_3	12 - 14	12	2
Fe_2O_3	0.25	0.35	-
TiO_2	-	0.5	-
F	0.1 – 0.2	0.45 (as F_2)	-

For each of these glasses, two batches with different borate raw materials were employed, and the batch compositions were calculated to prepare the same final glass (compositions indicated in Table 1). For each of these glasses, one batch was prepared using a hydrated borate raw material, and the other batch was prepared with an anhydrous borate (all the other raw materials were kept similar).

The low-κ and the Duran® batches were prepared with (hydrated) boric acid H_3BO_3 or (anhydrous) boric oxide B_2O_3. The other raw materials used included sand, sodium-, lithium-potassium-, calcium- and magnesium carbonate, alumina, titanium dioxide and fluorspar (calcium fluorine).

The E-glass batches were prepared using (hydrated) colemanite $2CaO.3B_2O_3.5H_2O$ or burnt (dehydrated) colemanite $2CaO.3B_2O_3$. The burnt colemanite was prepared from regular colemanite, using a calcination method described in [5]. The other raw materials used included sand, limestone (or quick lime), dolomite, kaolin ($Al_2O_3.2SiO_2.2H_2O$) and fluorspar.

All glasses were melted in an oxy-fuel fired furnace atmosphere, with approx. 55 % water vapor in the atmosphere.

The results of the EGA for low-κ and the Duran® batches are presented in Figures 3 and 4, respectively. For each type of glass, the results obtained for the batches prepared with boric acid and with boric oxide are compared. Attention is focused only on the CO_2 releases. From these figures, it can be seen that the type of borate raw material selected has an impact on the CO_2 release behavior. Indeed, for both low-κ and Duran® glasses, the batches prepared with boric acid show earlier CO_2 releases, with a first peak measured between approx. 180°C and 300°C. The CO_2 releases (due to carbonate decomposition) occur at higher temperatures for the batches prepared with boric oxide.

In the case of the low-κ batches, the CO_2 release profile above 400°C is roughly similar for both sources of borate raw materials, though the release at higher temperatures is shifted towards lower temperatures when using boric acid, probably as a consequence of the early CO_2 release below 300°C. In the case of the Duran® batches, only the CO_2 release profile above 650°C appears similar for both types of borate raw materials. The release measured between 400°C and 650°C in the case of boric oxide has shifted to lower temperatures (below 300°C) for the batch with boric acid.

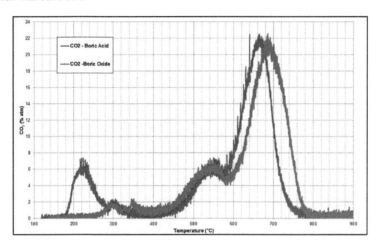

Figure 3. CO_2 evolution for the low-κ batches prepared with boric acid and boric oxide

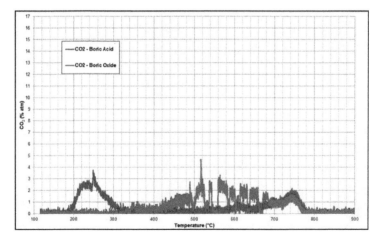

Figure 4. CO_2 evolution for the Duran® batches prepared with boric acid and boric oxide (NB: the CO_2 releases for the batch prepared with boric oxide shows a "spiky" profile due to partial blockage of the crucible outlet tube during the experiment)

The videos recorded during the melting of these batches with boric acid (which cannot be integrated into this manuscript) showed that in both cases, a molten phase was obtained at slightly lower temperatures (10°C to 20°C lower) as compared to the batches prepared with boric oxide. Therefore, the use of hydrated borate raw materials may lead to a faster melting-in of the batches. This confirms other results obtained for E-glass melts prepared using boric acid and boron oxide [6].

The CO_2 and SO_2 (sulfate-fined glass) gas evolution curves for the E-glass batches prepared with standard (hydrated) colemanite and burnt (anhydrous) colemanite are presented in Figure 5. For confidentiality reasons, the temperatures and amounts of gas released cannot be indicated for these experiments. The results do not show any early CO_2 release peak at lower temperatures, and the total emissions are only very slightly shifted towards lower temperatures for the batch with burnt colemanite. The SO_2 releases measured for both batches show some differences, both at lower and higher temperatures. These differences may be explained by different organic content in the batch (the burnt colemanite containing less organics after the calcination process) as the sulfate chemistry is influenced by the organic content in the batch [7], and/or a difference in the viscosity of the melt due to a difference in the water content in the melt.

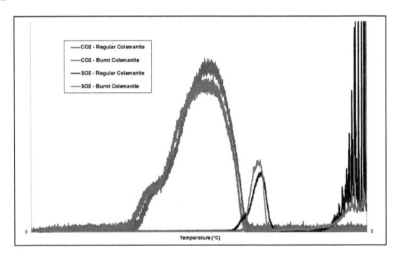

Figure 5. CO_2 and SO_2 evolution for the E-glass batches prepared with regular and burnt colemanite (units and values purposely skipped for confidentiality reasons)

The observation of the movie recorded during the melting of the two E-glass batches did not show substantial difference in the melting-in temperature for the batch prepared with hydrated colemanite. However, an impact was observed on the foaming behavior of the melts at higher temperatures. The foam height recorded at the same temperature is clearly different for the two batches, as can be seen in Figure 6. The batch prepared with burnt colemanite clearly shows a stronger foaming behavior, which corresponds to the higher SO_2 release at high temperatures shown in Figure 5.

The High Temperature Melting Observations and the Evolved Gas Analyses performed for these different alkali-lean glasses show that the choice of borate raw material, i.e. hydrated vs. anhydrous borate, can influence the reactions occurring in the batch, as well as the melting-in, fining and foaming behaviors of these glasses. It also shows that this influence can be different for different types of glasses.

In particular, the results on the low-dielectric and the Duran® glasses indicated that the use of boric acid leads to low-temperature reactions in the batch, provoking CO_2 releases at lower temperatures than for batches prepared with boric oxide. The observation of similarities in the release profile at higher temperatures indicates that only part of the carbonates is involved in these early stage reactions. In order to gain better understanding of these low temperature reactions, thermodynamic calculations have been conducted (results presented in the following section).

Figure 6. Difference in foam height at high temperatures for E-glass melts prepared with regular and burnt colemanite

THERMODYNAMIC CALCULATIONS

The results of the Evolved Gas Analysis showed that boric acid appears to promote low-temperature reactions, leading to release of CO_2 at low temperatures (below 300°C) in the low-κ and the Duran® batches, while this was not observed for batches prepared with boric oxide. In order to gain a better understanding of the mechanisms involved in these reactions, thermodynamic calculations were performed using the software FactSage (v. 6.3).

The reactions leading to CO_2 releases necessarily involve a carbonated raw material. Therefore, calculations were focused on the two main carbonated raw materials used in the batch investigated, namely sodium carbonates Na_2CO_3 for the Duran® batch and magnesium carbonate $MgCO_3$, for the low-κ glass. Other alkali- and alkaline-earth carbonates are expected to have relatively similar reaction paths, and the thermodynamic equilibria have not been calculated for these species.

FactSage was used to calculate the thermodynamic equilibrium resulting from the reaction of one mole of the borate raw material and one mole of the carbonate, at ambient pressure, and at different temperatures (every 10°C from 50°C to 500°C). The different possible products (gas, liquid, solid and slag phases) were calculated for the 4 following reactions:

- 1 mol H_3BO_3 + 1 mol $MgCO_3$
- 1 mol H_3BO_3 + 1 mol Na_2CO_3
- 1 mol B_2O_3 + 1 mol $MgCO_3$
- 1 mol B_2O_3 + 1 mol Na_2CO_3

The results of the calculations are presented in Figures 7 to 10. It is important to emphasize that these show thermodynamic calculation results, and the reactions kinetics are not taken into account. Therefore, the reactions identified are thermodynamically possible, but there is no indication whether or not they are kinetically possible (and thus whether or not they could be observed) at the timescale of the experiments performed.

When (hydrated) boric acid is used, the thermodynamic calculations indicate that this compound will first dehydrate to form metaboric acid and water below 100°C:

$$H_3BO_3 \rightarrow HBO_2 + H_2O$$

At higher temperatures, the metaboric acid can then react with the carbonates to form solid alkali- or alkaline-earth borates, with release of CO_2.

In the case of magnesium carbonate, the reactions occurring are:

$$2\,MgCO_3 + 2\,HBO_2 \rightarrow Mg_2B_2O_5\,(s) + 2\,CO_2\,(g) + H_2O \qquad \text{(above 100°C)}$$

Thus $\qquad 2\,MgCO_3 + 2\,H_3BO_3 \rightarrow Mg_2B_2O_5\,(s) + 2\,CO_2\,(g) + 3\,H_2O$

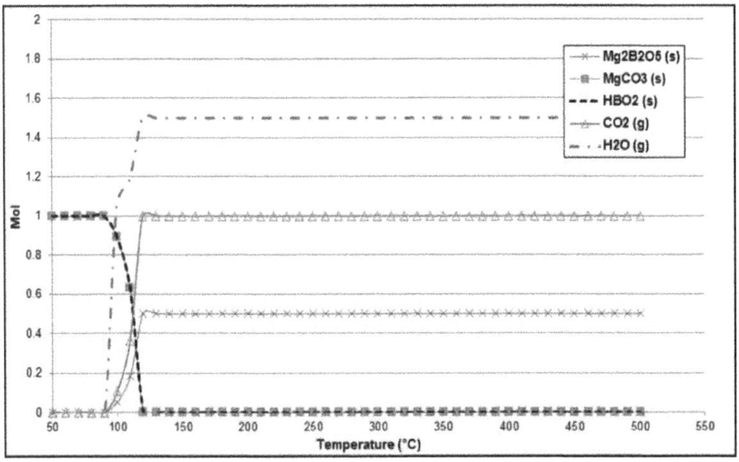

Figure 7. Thermodynamic calculations for reactions between $MgCO_3$ and H_3BO_3

In the case of sodium carbonate, the reactions occurring are

	$Na_2CO_3 + 6\ HBO_2$	$\rightarrow\ 2\ NaB_3O_5\ (s) + CO_2\ (g) + 3\ H_2O$	(above 100°C)
Followed by	$4\ NaB_3O_5 + Na_2CO_3$	$\rightarrow\ 3\ Na_2B_4O_7\ (s) + CO_2\ (g)$	(above 150°C)
Followed by	$Na_2B_4O_7 + Na_2CO_3$	$\rightarrow\ 4\ NaBO_2\ (s) + CO_2\ (g)$	(above 350°C)
Thus	*$Na_2CO_3 + 2\ H_3BO_3$*	*$\rightarrow\ 2\ NaBO_2\ (s) + CO_2 + 3\ H_2O$*	

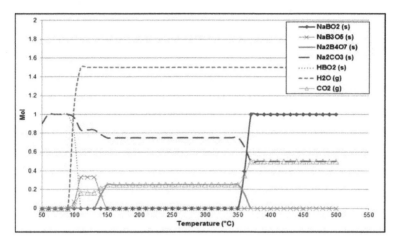

Figure 8. Thermodynamic calculations for reactions between Na_2CO_3 and H_3BO_3

When (anhydrous) boric oxide is used on the batch, a different set of reactions may occur, as described below. In the case of magnesium carbonate, the reactions occurring are:

	$MgCO_3 + 2\ B_2O_3$	$\rightarrow\ MgB_4O_7\ (s) + CO_2\ (g)$	(below 100°C)
Followed by	$3\ MgCO_3 + MgB_4O_7$	$\rightarrow\ 2\ Mg_2B_2O_5\ (s) + 3\ CO_2\ (g)$	(above 100°C)
Thus	*$2\ MgCO_3 + B_2O_3$*	*$\rightarrow\ Mg_2B_2O_5 + 2\ CO_2$*	

In the case of sodium carbonate, the reactions occurring are:

	$Na_2CO_3 + 3\ B_2O_3$	$\rightarrow\ 2\ NaB_3O_5\ (s) + CO_2\ (g)$	(below 180°C)
Followed by	$4\ NaB_3O_5 + Na_2CO_3$	$\rightarrow\ 3\ Na_2B_4O_7\ (s) + CO_2\ (g)$	(above 180°C)
Followed by	$Na_2CO_3 + Na_2B_4O_7$	$\rightarrow\ 4\ NaBO_2\ (s) + CO_2\ (g)$	(above 410°C)
Thus	*$Na_2CO_3 + B_2O_3$*	*$\rightarrow\ 2\ NaBO_2\ (s) + CO_2\ (g)$*	

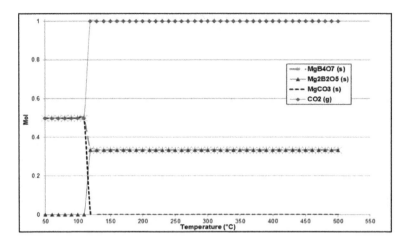

Figure 9. Thermodynamic calculations for reactions between $MgCO_3$ and B_2O_3

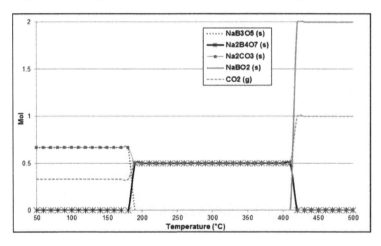

Figure 10. Thermodynamic calculations for reactions between Na_2CO_3 and B_2O_3

In both cases (anhydrous vs. hydrated borate raw materials), the products of reaction are the same, with production of gaseous CO_2 and solid-state $Mg_2B_2O_5$ for the reactions with magnesium carbonate and $NaBO_2$ for the reactions with sodium carbonate, respectively. However, the reactional paths leading to the formation of these solid compounds are different, with notably the formation of metaboric acid during the dehydration of the boric acid. This intermediate reaction could explain the observation of the early CO_2 releases during the EGA for the batches prepared with boric acid as compared to those prepared with boric oxide (see Figures 3 and 4). As explained previously, these thermodynamic calculations do not take into account the reaction kinetics. In fact, the dehydration step of the boric acid into water and metaboric acid may promote the subsequent reactions and lead to the early CO_2 releases. The water vapor released may promote the wetting of the batch and thus enhance the solid-state reactions leading to the formation of the alkali- or alkaline-earth borates, combined with the CO_2 release. The thermodynamic calculations suggest that the dehydration of the metaboric acid can occur at temperatures below 100°C. However, in practice, this reaction is known to occur at higher temperatures, at approximately 170°C. This temperature would match very well with the observation of the first CO_2 release at approx. 200°C for the batches with boric acid. It is likely that the solid state reactions leading to the formation of alkali- or alkaline-earth borates are not kinetically favorable in the absence of the dehydration process, explaining the absence of early CO_2 release in the case of the batches prepared with boric oxide.

Therefore, the use of boric acid in the batches instead of boric oxide leads to a shift of a part of the decarbonation reactions towards lower temperatures, via promotion of solid state reactions between the carbonates and intermediate alkali- or alkaline-earth borates. Additional experiments, such as in-situ XRD, would be necessary to identify the intermediate solid-state compounds formed and confirm these thermodynamic calculations. In addition, it is supposed that the different reactional paths between batches with boric acid and batches with boric oxide may, due to the formation of these solid-state intermediates at different temperatures, influence the evaporation of borate species from the batch at lower temperatures. Additional experimental work is required to investigate these reactions further.

WATER CONTENT IN THE GLASSES

The impact of the type of borate raw material (hydrated vs. anhydrous) on the final water content in the glass was investigated. The water content in the glass can be characterized by measuring the absorption of the –OH bands in the infrared around 2.8 µm. The water content in the glass can be calculated from the intensity of the –OH absorption band, using the following relation [8]:

$$H_2O \text{ content in glass } (ppm) = \frac{10^6 . M_{H2O} . {}^{10}\log(\frac{T_0}{T_{2.8}})}{d . \rho . e_{\rho,2.8}}$$

where d is the thickness of the sample (cm), T_0 and $T_{2.8}$ are the transmission at 2.5 µm (where no absorption by –OH groups occurs) and 2.8 µm, respectively, M_{H2O} is the molar mass of water, ρ is the density of the glass (in Kg.m-3) and $e_{\rho,2.8}$ is the practical linear extinction coefficient for dissolved water in the glass at a wavelength of 2.8 µm.

The practical linear extinction coefficient for dissolved water in the borosilicate glasses investigated being not known, the water contents have been calculated using an arbitrary value of $e_{\rho,2.8} = 40,000$ cm^{-1} (similar to that of soda-lime-silicate glasses). The results presented should therefore be considered purely as qualitative and not quantitative values, used only for comparison of the water content of glasses prepared with different borate raw materials. In other words, the values presented are only used to compare the water content in glasses prepared with anhydrous or with hydrated borates.

The glasses obtained from the HTMOS-EGA experiments were prepared in a "wet" atmosphere, containing approx. 55% water vapor. It is well known that the melting atmosphere also influences the final water content in the glass. In order to limit this impact and to focus on the effect of the borate raw materials, the glasses analyzed by FTIR were prepared separately, by melting the batches in platinum crucible under a dry (N_2) atmosphere. The glasses prepared were then cut into plane-parallel samples (thickness approx. 2.4 mm) and optically polished before analysis of the water content by FTIR spectrometry.

The measurements were carried out on the Duran® glasses (prepared with boric acid and boric oxide) and on the E-glass (prepared with regular colemanite and burnt colemanite). The batch compositions used were similar to those used for the HTMOS-EGA experiments. The recorded FTIR spectra for the E-glasses and for the Duran® glasses are presented in Figures 11 and 12, respectively. The calculated water contents are indicated in Table 2.

From the Figures 11 and 12, it can be seen that the –OH groups (dissolved water in the glass) do not absorb at the same wavelength in both glasses. The maximum of absorption is found at 2.82 μm and 2.77 μm for the E-glass and the Duran® glass, respectively. This difference results from the different structures of these glasses.

TABLE 2. Indicative water content (ppm) calculated for hydrated vs. anhydrous borate raw material in E-glass and Duran® glass

Borate raw material	E-glass	Duran® glass
Hydrated borate	215	506
Anhydrous borate	215	467

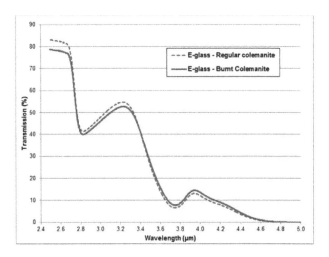

Figure 11. IR Transmission spectra for the E-glasses prepared with regular colemanite and burnt colemanite (thickness of the samples = 2.4 mm)

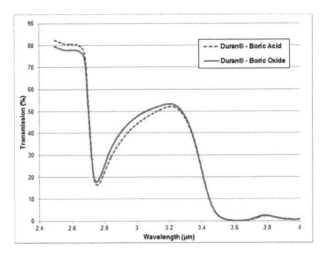

Figure 12. IR Transmission spectra for the Duran® glasses prepared with boric acid and boric oxide (thickness of the samples = 2.35 mm)

The results indicated in Table 2 show that the type of borate raw material may have an influence on the water content for the Duran® glass, with a lower water content measured for the glass prepared with anhydrous borate (boric oxide), as could be expected. In the case of E-glass the use of regular or burnt colemanite (hydrated vs. anhydrous) does not influence the final water content in the glass. However, in the case of the E-glasses, the batches were prepared using a significant amount of kaolin ($Al_2O_3.2SiO_2.2H_2O$) as a raw material, i.e. another hydrated raw material, while no other hydrated raw material was used to prepared Duran® glasses. In addition, the content in borate raw material used for preparation of the Duran® batches was significantly larger than for the E-glass batches, this glass having a higher boron content (see Table 1). Therefore, the influence of anhydrous vs hydrated colemanite in the case of E-glass may be masked by the influence of the other hydrated raw material (kaolin). The results from the Duran® glass show that, for glasses with high amounts of borate raw materials and without other hydrated raw materials, the use of hydrated borates may lead to higher water content in the glass.

CONCLUSIONS

In this paper, the influence of hydrated vs. anhydrous borate raw materials in different alkali-lean borosilicate glass batches has been investigated. The choice of either hydrated or anhydrous borate raw material has an impact on the glass melting process (batch reactions, melting-in, foaming), and this impact differs for the different borate raw materials and glasses investigated.

The melting-in behavior as well as the evolution of the gases released from the batches have been studied using the HTMOS-EGA setup. The results revealed that the use of regular colemanite as compared to burnt colemanite does not significantly impact the melting-in behavior, but has an influence on the foaming behavior. In the case of low-κ and Duran® glass batches, the use of (hydrated) boric acid leads to a slight reduction of the melting-in temperature.

In addition, the batches with boric acid show early CO_2 release below 300°C, which is not observed for the batches with boric oxide.

Thermodynamic calculations were used to explain this early CO_2 release. The results showed that the dehydration of boric acid may promote low-temperature solid-state reactions between the borate and the alkali- or alkaline-earth carbonated raw materials, to form solid alkali- or alkaline-earth borate compounds, combined with CO_2 release. Though thermodynamically possible for batches with boric oxide, in that case the reaction kinetics seem to be too low to be observed at the experimental timescale.

Finally, the water content in glasses prepared with hydrated and anhydrous borate raw materials was analyzed. These measurements revealed that the use of hydrated borate raw materials can lead to higher water content in the glass, for glasses prepared without other hydrated raw materials in the batch.

Therefore, the choice of hydrated vs. anhydrous borate raw materials in alkali-lean glasses can have an impact on the gas release, melting-in, fining (potentially foaming) and dissolved water content in the glass. The impact is more significant for glasses with higher boron content and when the batch does not contain other hydrated raw materials.

REFERENCES

1. M. Hubert and A.-J. Faber, *On the structural role of boron in borosilicate glasses.* Physics and Chemistry of Glasses (European Journal of Glass Science and Technology Part B), 2014. **55**(3): p. 136-158.
2. R.A. Smith, *Boron in glass and glass making.* Journal of Non-Crystalline Solids, 1986. **84**(1–3): p. 421-432.
3. A. Zamurs, S. Donthu, and S. Cook, *High Value Borates for the Glass Industry: Dehydrated Borates*, in *GlassTrend - DGG/HVG - NCNG Seminar on Alternative Raw Materials and Advanced Batch Pretreatment for Glass Melting.* 2012: Eindhoven, The Netherlands.
4. A.D. Puri, *Choice between anhydrous borax and borax hydrates.* Glass Technology, 2000. **41**(6): p. 174-176.
5. S. Sener and G. Ozbayoglu, *Separation of ulexite from colemanite by calcination.* Minerals Engineering, 1995. **8**(6): p. 697-704.
6. A. Zamurs, *The Use of Hot Stage Microscopy to Observe and Analyze Glass Formation Behavior*, in *Glass & Optical Materials Division and Deutsche Glastechnische Gesellschaft - ACerS GOMD–DGG Joint Meeting.* 2015: Miami (FL), USA.
7. R. Beerkens and J. van-der-Schaaf, *Gas release and foam formation during melting and fining of glasses.* Journal of the American Ceramic Society, 2006. **89**(1): p. 24-35.
8. NCNG - Multiple authors, *Glass Technology Course Textbook 2012 - Part 1.*

GLASS CULLET: IMPACT OF COLOR SORTING ON GLASS REDOX STATE

Stefano Ceola, Nicola Favaro, and Antonio Daneo
Stazione Sperimentale del Vetro, Via Briati 10, 30141 Murano – Venezia (VE) Italy

SUMMARY

This paper describes some common problems in the use of cullet as a secondary raw material for the glass industry. First an overview of the use of cullet at EU level will be presented, along with its advantages and disadvantages. Then the paper focuses on the complex European cullet collection and treatment system.

After an overview of the most common issues derived from the use of cullet, special attention is drawn to cullet contributions to possible instabilities in the final redox ratio of the produced glass. A new analytical methodology is described, the Inorganic Redox (I-Redox), to check the quality and stability of the produced cullet.

The determination of the redox ratio is described theoretically and experimentally as it is usually performed in glass industry as a quality check tool, then the I-Redox is described and discussed as a new tool for the determination of the redox contribution of cullet in the batch and as a parameter for monitoring cullet redox variability.

SCOPE OF THE PAPER

One of the main parameters of study and concern in glass industry is the redox state of the glass. It drives the sulfur/sulfate solubility and the color of the glass [2], and its stability is of key importance in the frame of quality control of the produced glass.

In current practice cullet represents a large part of a batch charge, reaching up to 90% of the batch in some European glass processes. Its use gives many energy and environmental advantages, and it has become an irreplaceable component in the majority of large-scale glass production.

Glass industry has extensive cumulative experience in the use of cullet, overcoming several difficulties occurred since its firsts use in the eighties [3], by finding suitable technological solutions or working with to cullet suppliers for a cleaner and more stable product.

This paper describes a new analytical tool for the control and evaluation of cullet effects on the glass final redox state. A new method developed by SSV to determine the Inorganic Redox is described and its application on some real cases illustrated.

CONTEXT OF THE RESEARCH

Commercial Recycled glass, or cullet is currently one of the main raw materials for the glass industry in Europe. More than 60% of glass cullet is recycled to produce new glass packaging, with some furnaces producing green color glasses having more than a 90 % recycling rate (see Figure 1).

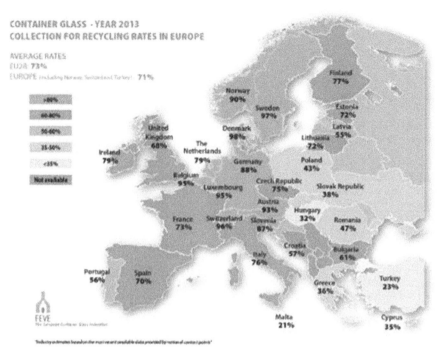

Figure 1. Container Glass Recycling Rates in Europe (2013)

Use of cullet is very advantageous for the glass industry. The main pro's and con's are reported in Table 1.

Table 1. Advantages vs. Disadvantages of Cullet Use in Glass Industry

Advantages	Disadvantages
Less raw materials use.	Energy consumption (cullet processing).
Less CO_2 and thus lower meelting energy for processing raw materials.	Indirect CO_2 emission (cullet collection).
Less direct CO_2 emission from raw materials provides lower transportation costs.	Glass quality issues (redox, metals etc.).
Less energy for melting.	Higher abrasive wear on some systems (?)
Less CO_2 emissions from melting.	
Less indirect energy consumption and CO_2 emissions. (extractions).	

Main source of cullet are recyclables from separate waste collection systems. Post-consumer dry recyclables are collected primarily in two main ways in Europe: mono-material collection (optionally with color differentiation), or mixed with other dry recyclables (multi-material collection; plastic, mixed color glass, cans). This means that glass industry in each country has a different approach to possible issues arising from cullet use.

In the case of multi-material collection systems, the material collected by the municipal collection centers is composed of glass along with plastic, aluminum, metal scrap, stones, ceramic, porcelain, wood, and (Figure 2).

Figure 2. Example of Multi-Material Collection System

Treatment process

The treatment process consists of cleaning of the waste from any contaminants such as plastic, ceramic, stones, metals in order to obtain an almost clean cullet. This is not only required for the recycling of the cullet in the glass furnace, but also to abide by the End of Waste European regulation. This regulation establishes the criteria that define how a material ceases to be waste, in order to be stored, shipped and used as secondary raw material.

The treatment process consists of a series of steps, conceived to efficiently clean and remove any non-glassy material from the recyclables. For example, to clean and remove aluminum and non-ferrous scrap metal, the material passes over an electric grid, which efficiently separates aluminum scraps from glass. The advantage of getting rid of aluminum couples with the recovery of aluminum itself, which is another valuable recyclable source for metal industry.

The details of the process may vary between different companies: the schematic in Figure 3 is only a general description of one possible process designed to make clean cullet. For some companies the order of the operation can vary, depending on the type of incoming material.

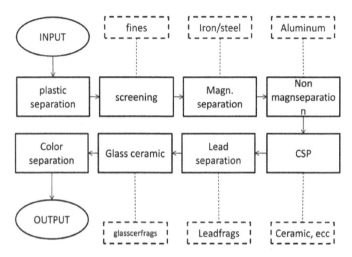

Figure 3. Schematic of a General Description of a Process

The input material comes from the municipal waste dry recyclables, and the output can either be mixed cullet, or flint cullet produced by color separation plus the remaining cullet. In fact, color separation is usually applied to mixed cullet to get white or half-white cullet, and the rest is a mixed cullet with decreased less flint component. If managed correctly, such treatment results in a final mixed cullet product that is virtually clean from ceramic, stone, plastic and metallic contaminants (Figure 4).

Figure 4. Final Mixed Cullet Product

The introduction of color sorting machines has a positive impact on the amount of cullet recycled, however some unexpected contraindications have been registered. The redox of the cullet fed depends by the relative amounts of the different colors. Usual mixed cullet containing flint glass is no longer widely available in the market, and the colored cullet obtained after flint cullet separation is a mixture of green and amber cullet. This change in color composition leads to a different product. For example, when used in the production of colored glass, small variation can modify the redox of the batch, introducing instability on the glass color, especially in case of reduced glasses (amber and UVAG).

Sorting machines

During cullet treatment or processing, sorting machines exploiting different detection systems are able to remove fragments of undesirable contaminants.

There are in general four types of detection systems:

- Visible-Near Infrared detector: able to remove glass ceramic;
- Visible light detection: able to remove opaque material (CSP) or suitable for color separation;
- UV Fluorescence detection: for the removal of lead glass fragments;
- X-Ray Fluorescence detection: for the removal of material with a specific composition, especially useful for lead glass fragments

These machines in general work continuous by letting the cullet flow through the detection system, which drives the subsequent array of air blows, activated when the detector recognizes the contaminating material (Figure 5).

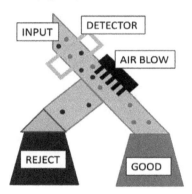

Figure 5. Detector Configuration

In actual practice, even the highest quality cullet is not completely free from such contaminants, which could can potentially cause severe issues as stone and metallic inclusion in the final product.

For varying reasons the efficiency of removal of these machines are less than 100%, influenced by:

- Condition of material during treatment: the capability of a system to recognize the undesired material depends on the degree of cleanliness of the glass (the amount of contamination attached to the glass fragments), or the humidity level of the material itself;

- Type of material: the supply of waste to be processed is not always constant, since it may come from different geographical places, and the calibration of the sorting machine must be optimized accordingly;
- Air blow: intrinsically, the air blow is not selective and it does not remove only the undesired fragments, but also the glass fragments in close proximity to the contaminants;
- Stream effect: since the air blow is not instantaneous but has a certain finite duration, during this time other fragments in the tail of the target fragment can be removed even with the best synchronization.

Because of these effects, the rejected material can contain "good" glass, while the clean glass can still contain some contaminants. In the case of color separation, a mixed cullet is treated to obtain almost clean white cullet, and a mixed cullet enriched in the colored component. For the cullet producer, a white cullet free from colored glass fragments has a high value in the market, and color separation produces a colored cullet variably contaminated by flint glass fragments. It is this variability, because of the abovementioned effect, that could potentially give an unstable product, from the point of view of redox contribution to the final glass.

ANALYTICAL METHODOLOGY

Description of the methodology: The following paragraph describes the procedure employed to determine the cullet redox ratio, that is the I-Redox, starting from the material properly sampled in the glass plant.

A sample of at least 50 kg of cullet is ground to a size below 3mm. The ground materials then evenly divided by a sampler divider, to obtain a sample of 1 kg and submitted for melting. The melting is performed by first keeping the material in a ceramic crucible for three hours at 550°C, in order to burn out the remaining organic materials (i.e. wood, paper, or plastic).

After organic removal, the cullet is further heated and melted at 1350°C, until a bubble-free glass is obtained, then cast and finally annealed at 530°C. The sample thus obtained is cut and polished to obtain suitable samples for XRF analysis and UV-Visible spectrophotometric analysis.

Experimental determination of the redox ratio: Redox ratio is a composite value, which is expressed by the following formula:

$$\% \text{ REDOX} = \text{Fe(II)} / \text{Fe}_2\text{O}_3\text{(tot)}$$

The value of the reduced iron, Fe(II), is expressed as FeO and determined by Visible-Near Infrared (Vis-NIR) spectrophotometry. A glass sample of suitable dimension is placed in a spectrophotometer and its Vis-NIR spectra is registered. The signal at 1050 nm is elaborated by a procedure described elsewhere [1], by which the concentration of FeO in the glass is calculated.

The Total Iron is measured by X-Ray Fluorescence Spectrometry (WDS-XRF), and expressed as Fe_2O_3. The results are calculated after a comparison with a calibration curve for each chemical species pertinent to the analysis.

Calculation of the analytical uncertainty: The calculation of the analytical uncertainty for the I-Redox takes into account the analytical uncertainty of the XRF value for the total iron, expressed as Fe_2O_3, and the uncertainty of FeO value determined via UV-Vis spectrophotometry. The calculation of the analytical uncertainty starts from the evidence that the two values considered are uncorrelated. In this case, we can apply the formula:

$$u(y) = \sqrt{\sum_{i=1}^{n}\left[\left(\frac{\partial y}{\partial x_i}\right)^2 \bullet u^2(\bar{x}_i)\right]}$$

where "y" is the redox ratio, "xi" are the value for FeO and Fe_2O_3, and "u" represents the experimental uncertainty.

From the evaluation of the experimental errors for each test, XRF and UV-NIR spectroscopy, estimated respectively at 50 ppm and 50 ppm, the analytical uncertainty for reduced glass is 1.2% – 1.8% of redox unity. The following formula gives the espression for u(y) in the case of the redox value, where A is the concentration of FeO (Vis-NIR) and B is for the concentration of Fe_2O_3 (XRF).

$$\frac{1}{B}\sqrt{u_A^2 + \frac{A^2}{B^2}u_B^2}$$

RESULTS AND DISCUSSION

Validation of the Methodology– Repeatability
The methodology described above has been applied to the same pile of cullet by measuring redox of a series of samples, with each sample gathered using the same procedure. The results in Table 2 show the calculated standard deviation for each measurement and for the calculated redox. The variability in the same lot of material is quite low, below 10% of the average value.

Table 2. Redox Data with Calculated Standard Deviation

	1	2	3	4	media	dev
Fe_2O_3	0.30	0.30	0.29	0.31	0.30	0.006
FeO	0.092	0.089	0.094	0.10	0.093	0.004
Redox %	34	33.2	35.9	36.2	34.8	1.3

Methodology applied to cullet monitoring - Variability
In order to test the variability of the I-Redox, the methodology was applied to a series of cullet samples coming from the cullet used in an operating, one sample for each day of use. The results are shown in Table 3 and in Figure 6.

Table 3. Redox Data with Calculated Standard Deviation

	1	2	3	4	5	6	7	Std dev
FeO	0.22	0.2	0.23	0.22	0.23	0.18	0.17	0.022
Fe₂O₃	0.37	0.39	0.38	0.37	0.38	0.36	0.36	0.010
Redox	66.1%	57.0%	67.3%	66.1%	67.3%	55.6%	52.5%	5.9%

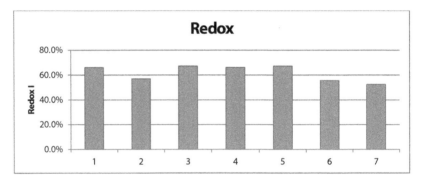

Figure 6. Redox Data

APPLICATIONS

The methodology for determining the Inorganic Redox can be used in the monitoring of the contribution of cullet to the redox of the batch. Actually, in many plants the contribution of organic contaminants is monitored daily with Loss On Ignition (LOI) testing. The organics participate in the chemical reaction during batch melting, and the corresponding formulation of the reducing agent in the batch is adjusted according to the tested LOI.

In Figure 7 the daily variability of LOI from the same cullet pile is shown. The standard deviation is 0.23%, with an average value of 0.55%, a maximum value of -0.95%, and a minimum value of 0.34%. Ina 300 ton Furnace using 60% cullet, the inorganic content coming from 180 tons of cullet could range between 612 to 1710 kg, to be compensated with the compatible amount of reducing agent in the batch.

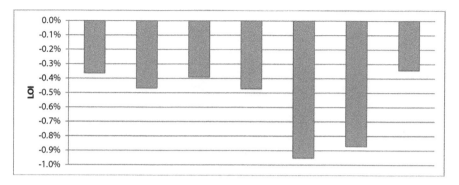

Figure 7. Daily Variability of LOI

The knowledge of the LOI does not prevent the arising of issue related to color and redox instabilities, especially for reduced glass. A different approach is proposed for the control and correction of the redox variability of cullet.

The possibility to know the redox ratio of the incoming cullet will be of help for the control of color of the final container. If the analysis shows an important variation of the Inorganic Redox of the cullet, corrective action for this can be undertaken as well, in terms of reducing agent, or iron supply to the batch.

We can simulate a typical situation for a furnace producing amber glass. A possible recipe is shown in Table 4.

Table 4. Recipe for Simulation

Component	% in weight
Internal cullet (Amber glass)	20
External (mixed) cullet	60
Batch	20

In this case, the amber glass produced has a redox of 79%, as Fe(II)/Fe(tot), with Fe_2O_3(total) of 0.51% and FeO of 0.36%. The variability of the cullet in use is shown in Table 3, with a standard deviation of 5.8%. With external cullet making up 63% of the total glass produced (Glass yield is more than 95%), the contribution to the standard deviation is 3.7%, more than double the analytical uncertainty. This variability due to the cullet can influence the quality of the produced glass, because redox could range easily from 77 to 82%. By having knowledge of the Inorganic Redox variability of the incoming cullet, it is possible to implement an integrated approach, taking into account the response of the system (furnace) to external solicitation, by using a Simulink® model.

Dynamic model

Simulink® is a Matlab® tool developed for the simulation of closed system, with the possibility of analyzing the response of the system after a variation of the boundary condition during time, taking into account all the other parameters (Figure 8).

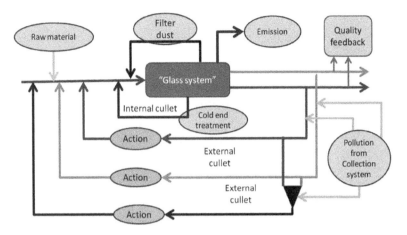

Figure 8. Glass System Developed using Matlab®

In the case discussed in this paper the system is a glass furnace, which is modeled as a mixer with its Residence Time Distribution (RTD) behavior, known by the furnace manager. The system is provided with an input (batch and cullet) and output, which is the produced glass. Each input and output is characterized by one or more parameters, linked to the type of investigation implemented. As simple example, if the scope of the investigation is to find the effect of the lowering of soda in the batch, the input parameter is the quantity of soda, and the output can be the CT (Cooling Time of the glass) or the Relative Machine Speed (RMS). In this way, the glass technology can simulate the effect of soda variation on the output parameters, both quantitatively and its impact over time, such as predicting the time at which it stabilizes. If other parameters are defined (CO_2 emission or energy consumption, for example) the evaluation is enriched by additional, useful information. As is depicted in Figure 9, the system allows different scenarios to be compared and evaluated.

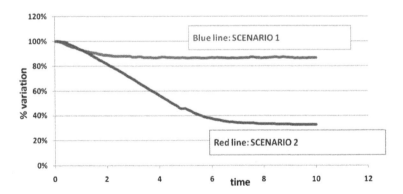

Figure 9. Comparison of Different Scenarios

In the case discussed in this paper, the input parameter could be the redox variability of cullet, and the output parameter the quality of the final container, as redox ratio and/or color. In this simple form, it is possible to evaluate the impact of the variation in time, taking into account the reaction time of the furnace (represented by the RTD). From this starting point, the system can be enriched with other data, such as the variability arising from use of mixed cullet coming from color separation, introducing also the possibility to change the batch redox. This improvement in the model could eventually lead to refining the knowledge of the system, enabling simulation of possible corrective actions beforehand. The next phase of the work will be the development of such predictive model and the application at industrial scale.

CONCLUSION

In this paper, the authors propose a new approach for monitoring and control of glass cullet. This methodology applies to cullet cleaned from contaminants, such as plastic, metals and CSP stones, also called Ready-to-Furnace Cullet. The main source of cullet is Separate Waste Collection System, as is currently widely applied in Europe, with some difference from country to country. For some countries, cullet is collected along with plastic and metals. Due to common misunderstanding of the collection rules by the citizens, glass recycled in this way may also contain ceramic, stones and porcelain. This material, if present in cullet, will eventually be the source of critical defects in the final product of glass industry, especially container glass.

In order to obtain cullet as free as possible from potential sources of problems in the final product, waste treatment companies apply a process for cleaning the material and sorting out all the possible contaminants. This process eventually leads to a material almost free from contaminants, which can be safely used in the glass furnace, at rates of up to 90% in the batch charge. Even the cleanest cullet may still contribute significant instability to the glass process, as LOI variation, and also by contributing variation coming from its compositions, that is the redox ratio. A standard methodology is proposed for the determination of the redox ratio of cullet, called Inorganic Redox. This parameter can characterize the effect of cullet on the final redox state of the produced container, especially in the evaluation of its instability.

This paper proposes an approach based on a dynamic simulation, where a careful setup of the model and a proper choice of the parameters could eventually lead to a reliable simulation of the instability on the final product, and the comparison of the different effect of the possible corrective actions. Further studies are underwayfor the application of this approach to different scenarios, especially regarding reliable forecast about input variability and analysis of side effect after input change.

REFERENCES

[1]   "Fast Spectrophotometric Determination of Fe^{3+} and Fe^{2+} in commercial glass", A. Daneo, P. Polato, M. L. Scandellari, M. Verità, Proceedings of XVII International Congress of Glass, vol. 3, pg 83-88, Chinese Ceramic Society Beijing (1995).

[2]   "Diffusion and redox Reactions of Sulfur in silicate Melts", H. Beherens, J. Stelling, Review in Mineralogy and Geochemistry, Vol. 73, pg 79-103, (2011).

[3]   "Problems related to the use of Cullet and to the reuse of dusts from Fumes Treatment Plants", B. M. Scalet, International Glass Journal, No. 87, pg 65-69, (1996).

Combustion, Refractories, and Sensors

OXYGEN-SOLID FUEL COMBUSTION IN GLASS MELTING FURNACES

Dr. Mark D. D'Agostini, Dr. Jinghong Wang, Mr. Juping Zhao
Air Products and Chemicals

ABSTRACT

With prices of conventional fossil fuels such as natural gas and fuel oil projected to remain relatively high in several regions of the world while the global demand for glass production is comparatively low, many glass manufacturers, particularly in Asia and Latin America, are viewing petroleum coke (petcoke) and, secondarily, coal as potentially viable, low-cost, alternative fuels. There are, however, certain challenges and risks associated with solid fuel utilization for glass melting that need to be understood prior to commercial adaptation. And while petcoke / coal combustion may not be an acceptable choice in every case, Air Products' experience and know-how suggest that enhancement of solid fuel combustion with oxygen will broaden its successful adaptation in glass melting applications relative to the use of air-fuel combustion. This paper explains the challenges and risks of petcoke and coal combustion for glass melting, and via laboratory and field data, highlights the benefit that oxygen enrichment can bring.

PROPERTIES OF PETCOKE AND COAL

Petcoke is a solid, essentially carbonaceous, by-product of crude oil refining. As such, it is not surprising that petcoke chemical properties are quite similar to heavy fuel oil (HFO). This is illustrated in Tables I and II, where relevant property ranges are shown for the two types of

Table I
Major Constituents of Typical Fuel Grade Petcoke and Heavy Fuel Oil
(wt%, dry basis)

Constituent	Fuel Grade Petcoke	Heavy Fuel Oil
Carbon	85 - 90	83 - 88
Hydrogen	3 - 6	10.5 – 11.0
Nitrogen	0.1 – 2.0	0.15 – 0.40
Oxygen	0 – 1	0.35 – 0.40
Sulfur	4 – 7	2 – 4
Ash	0.1 – 0.5	0.04 – 0.20
Moisture	0.5 – 10	0.3

fuel. Although properties vary substantially with the crude oil source and refinement method, it is evident that, apart from marginally higher sulfur and ash content, (fuel grade) petcoke and heavy fuel oil are, chemically, very similar. Moreover, comparison of petcoke with bituminous coal (Table II) reveals that petcoke generally has much lower volatile matter content and ash content. The low ash content makes petcoke more attractive for glass melting than coal, since it reduces the risk of ash mineral-related glass contamination and defects, and also lowers particulate emissions. However, as subsequently explained, the low volatile matter represents a principal challenge in the effective utilization of petcoke for glass melting.

Table II
Trace Metals of Typical Fuel Grade Petcoke and Heavy Fuel Oil
(ppmw of ash)

Fly Ash Metal	Fuel Grade Petcoke	Heavy Fuel Oil
Aluminum, as Al_2O_3	40,000 – 70,000	3500 – 123,000
Calcium, as CaO	10,000 – 155,000	5300 – 25,700
Chromium	10 – 100	48 - 4390
Iron, as Fe_2O_3	10,000 – 70,000	9500 – 488,000
Manganese	70 – 300	64 - 1170
Magnesium	15,000 – 24,000	2300 – 211,000
Molybdenum	10 - 20	22 – 2860
Nickel	1200 – 7500	820 – 41,600
Potassium, as K_2O	1000 – 12,000	400 – 80,600
Silicon, as SiO_2	12,000 – 350,000	6000 – 216,000
Sodium, as Na_2O	1800 – 14,000	1632 - 2480
Vanadium, as V_2O_5	500 – 400,000	2200 – 112,000

COMBUSTION CHALLENGES OF PETCOKE

Of all the properties highlighted in Table III, volatile matter, which comprises several light and heavy hydrocarbons, is the one with the most significant effect on practical petcoke combustion. In particular, the lower the volatile matter, the more difficult the solid fuel is to

Table III
Trace Metals of Typical Fuel Grade Petcoke and Heavy Fuel Oil
(ppmw of ash)

	Fuel Grade Petcoke	Bituminous Coal
Sulfur	4 – 7	2 – 4
Ash	0.1 – 0.5	6 – 12
Volatiles	8 – 12	30 - 40

ignite. Data obtained by Air Products and presented in Figure 1 indicate, for example, that when ignited in room temperature air, petcoke having 10 wt% volatile matter has an ignition energy that is several orders of magnitude higher than a common bituminous coal having 35 wt% volatiles. This more-difficult-to-achieve ignition of petcoke relative to bituminous coal renders its efficient combustion with air difficult to achieve in many applications such as glass melting furnaces, where relatively short residence times are available for the solid fuel to heat up, ignite and burn.

The most commonly applied remedial measure for addressing combustion residence time limitations is fine pulverization of the fuel. In coal combustion, for example, it is generally accepted that grind sizes of the order of 70 wt% or greater of the solid fuel passing through a 200 mesh screen (aperture size of approximately 75 microns) are required for efficient combustion with entrained-flow burners. Fine pulverization is indeed essential in applications having limited residence time, of which glass melting is one, yet industry experience suggests that fine pulverization alone is generally not sufficient, particularly in the case of petcoke. This is because while combustion rates are increased for finer particles (due to more abundant surface area), the added surface area has little effect on ignition energy[1]. However, as seen in Figure 1, when the aforementioned petcoke and bituminous coal ignition energy data are extended to include the effect of oxygen enrichment on the ignition atmosphere, a dramatic reduction in ignition energy is thereby obtained. And although an oxygen atmosphere of 50 mol% (balance N2) is needed to lower the petcoke ignition energy to that of the air-bituminous coal mixture, much lower levels of oxygen enrichment, properly applied, can have a substantial beneficial effect on combustion kinetics.

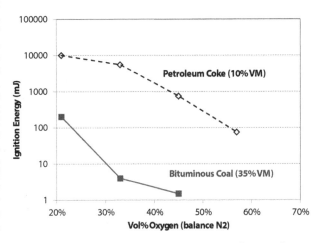

Figure 1. Ignition energy vs oxygen concentration for typical petcoke and bituminous coal

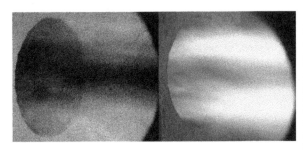

Figure 2. Photos of petcoke flame. A (left): Air-petcoke flame with oxygen staging. B (right): Oxygen-enriched air-petcoke flame using SF Port Injector.

To illustrate this point for an application with commercial relevance to glass melting, several industrial-scale solid fuel injection nozzles were tested in Air Products Clean Energy Laboratory (CEL), a pilot-scale, multi-fuel facility capable of firing rates up to 20 MMBtu/hr. Key results obtained during the development are highlighted in Figures 2 and 3. The photos in Figure 2, were taken in the CEL for two different simulations of oxygen enhanced combustion of an air-petcoke stream. Both scenarios represent identical conditions in firing rate, transportation air flow rate and overall flame stoichiometry, but differ as to the manner in which oxygen is introduced into the flame. In photo A, transportation air alone, amounting to approximately 10% of the total stoichiometric requirement, conveys the petcoke into a furnace, while the balance of oxygen required for complete combustion is lanced in a parallel stream immediately beneath the injector. By contrast, photo B shows the flame produced by Air Products' CLEANFIRE® SF Port Injector (SF for solid fuel), wherein a small fraction of the combustion oxygen is mixed with petcoke within the injector in a proprietary manner, while the balance of oxygen for complete combustion is lanced within the same parallel stream beneath the flame. The early ignition of petcoke with the SF Injector, leading to an "attached" flame, is clearly evident. We have repeatedly observed how, relative to a "lifted" or "detached" flame (as in photo A), the attached flame stabilizes combustion and substantially reduces flame pulsations and furnace pressure fluctuations, while increasing carbon burnout and heat transfer from the flame to its surroundings. Regarding heat transfer to the surroundings, the crown temperature along the furnace axis was recorded during operation with both the air-fuel and SF injectors. Results are plotted in Figure 3 as a function of distance from the hot face of the injector wall. Note how the furnace crown temperatures for the SF injector are systematically higher across the furnace length by between 50 and 150 degrees F, simply due to the early ignition achieved through optimized mixing of a small fraction of the combustion oxygen within the SF injector. We reiterate that the same total flow rate of oxygen was employed for both cases; only the method of introduction was different. The conclusions to be derived from this illustrated are that a) there is a tremendous potential for enhanced solid fuel energy utilization with oxygen, and that b) the oxygen-solid fuel mixing processes must be well controlled to realize the maximum beneficial effect.

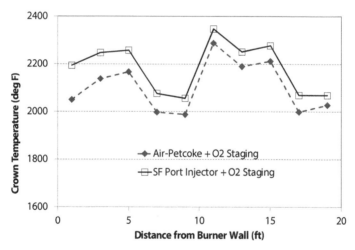

Figure 3. Lab data showing crown temperature vs distance from burner wall for air-petcoke and SF Port Injector tests.

The lab results for the SF injector vs straight oxygen lancing holds further implication to the effectiveness of port firing of petcoke in a regenerative glass melting furnace. Preliminary results for such an application are provided in a later section of this paper within one of our case studies.

IMPORTANCE OF SOLID FUEL CHEMISTRY

Ash from both petcoke and coal contain numerous metals and minerals that have the potential to influence glass quality, color and refractory life. Among these are refractory oxides such as alumina, calcium oxide (lime) and magnesia; metals including sulfur, vanadium and nickel, and naturally carbon, which in and of itself is a strong reducing agent. Non silica-based refractory elements, for example, is known to lead to stones and knots in glass, while inclusions of nickel sulfide, NiS, have been implicated in relation to spontaneous breakage in flat, tempered glass [2]. As such, it is obviously a concern with petcoke firing in float furnaces in particular. Nickel can also impart a color to the glass that varies with the composition of the glass matrix.

Vanadium is a trace metal that is known to react with other commonly occurring ash metals such as nickel, iron and sodium to form high melting point compounds. Silica, alumina and calcium have also been found to adhere to these compounds (once deposited) as separate species[3]. The deposited vanadium compounds, particularly vanadium pentoxide, V_2O_5, and salts such as the various sodium – vanadium compounds, have substantial potential to attack refractory and foul regenerator flow passages. V_2O_5 is a known catalyst with peak activity in the range of $950 - 1300$ degrees F [4] and a melting point of approximately 1150 degrees F. Accordingly, in regenerative glass furnaces, deposition and corrosive attack induced by vanadium is particularly prone to occur in regenerators. Finally, vanadium ions in a highly oxidized state are known to affect glass color, typically by imparting either a greenish or brownish tint [5].

A key concern of sulfur in solid fuel is that, via reaction with alkali metals it forms alkali sulfates which can attack and weaken refractory structures [6]. For example, sodium sulfate, Na_2SO_4, which melts at 1623 degrees F is can penetrate and lead to progressive deterioration of alumina-silicate refractories. Of further concern are emissions of sulfur dioxide (SO_2), which is a principle contributor to acid rain, and sulfur trioxide (SO_3) which combines with water vapor to form sulfuric acid, H_2SO_4. The acid then condenses from flue gas in the cold end of the flue gas ductwork, while also forming a condensate mist that can add a visible bluish tint to the furnace exhaust plume. Flue gas scrubbers may be required depending upon regional emissions regulations.

For petcoke in particular, carbon can be the most abundant element in fly ash, and its strongly reducing nature can effect glass redox state [7]. Our experience is that poor petcoke combustion can in practice lead to ash that contains over 90% carbon by weight. Apart from reducing fuel efficiency, which is naturally undesirable, high unburned carbon-in-ash levels produce larger, heavier fly ash particles which are more prone to find their way into both the glass melt and regenerator checker packs. Since, the above mentioned minerals and metals are also constituents of the fly ash, poor combustion will dramatically increase the any deleterious effect of these aforementioned species.

Based on these perspectives, it is clear that the benefits of oxygen enrichment in solid fuel combustion extend beyond fuel efficiency and into the realm of capital and maintenance costs and glass quality.

MATERIAL HANDLING

Combustion systems firing solid fuels can be categorized as directly or indirectly fired. Directly fired systems are those in which coarse fuel is metered and introduced into a grinding mill which, after completion of the grinding process, delivers the pulverized fuel with conveying air into one or more transport lines that carry fuel into the burners. The indirect firing system differs from this in that the pulverization step is decoupled from the transport of the fuel to the burner.

The decoupling occurs by short or long term storage of the pulverized fuel in a bin, hopper or other storage container. The indirect firing system is generally preferred for glass melting for at least the following two reasons:

1. For an indirectly fired system, interruptions in the grinding process, which are not uncommon, will not cause an interruption in the fuel being delivered to the furnace.
2. Pulverizers generally require more cool air than is strictly required for lean phase particle transport. Hence, burners in indirectly fired systems are generally smaller, and combustion is less diluted with cool transport air than for directly fired systems.

It is further important to mention that due to its low volatility, long-term storage of pulverized petcoke is generally not plagued by self-heating and spontaneous combustion. Thus, whether or not the pulverization process is carried out onsite or at remote location, storage of the pulverized petcoke onsite is a safe and viable option.

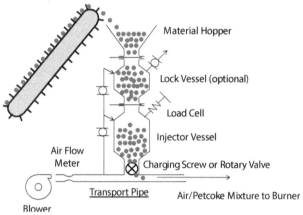

Figure 4. Typical material handling and transportation system for indirect firing.

Assuming the indirectly fired system is used, the basic material handling equipment package consists of a storage hopper, dust collection system, injector vessel, metering device and transport air blower. One common configuration illustrating the packaging of this equipment, along with auxiliary piping hardware is presented in Figure 4. (Note that while dust collection in the vicinity of the material handling system is strongly recommended, it is omitted from Figure 4 for the sake of brevity.) Many specific design variations are possible within the basic framework of this system. Perhaps the most critical of these relates to the maximum transportation air pressure required at the point of injection of solid fuel into the transportation air line. This is due to the cascading effect that pressure at this initial injection point has upon critical design and operating factors such as air blower / compressor cost and power requirement, material flow stability, dust control, injector vessel design and the need for a pressure isolation vessel. Concerning this latter point, Figure 4 depicts a so-called lock vessel, labeled as optional, which is frequently used for systems where a large pressure increase across the charging screw would otherwise be required. This arrangement provides a buffer between the atmospheric pressure hopper and higher pressure injector vessel, facilitating smooth and continuous material feeding across the charging screw.

In so far as it is highly desirable to maintain the lowest pressure possible at the point of discharge from the charging screw to the transportation line, selection of the transportation air flow rate and transport line size are of considerable importance. Transportation line velocities of between 15 and 30 m/sec are typically acceptable; depending upon particle size, shape and density. Velocity below 15 m/sec can lead to particle drop out and line plugging, while velocity above 30 m/sec will substantially increase both pressure drop and line erosion phenomena. Simultaneously, fuel to transportation air mass flow rate ratios between nominally 1 and 10 are generally preferred for glass melting. A ratio lower than 1 can be too lean for combustion applications, and may promote combustion instabilities, while also diluting the balance of oxidizer mixed at the burner, which will generally be either be hot air or oxygen. Conversely, fuel to transportation air ratios significantly above 10 can bring the onset of unsteady, chugging flow and higher pressure drop associated with dense phase transport. It is emphasized that the stated design ranges for the transportation line are intended for estimation only, and a more precise assessment should include the flow characteristics of the solid fuel particulate as well as details of the transportation line layout.

Concerning integration of the solid fuel handling system with the furnace, the most significant factor is whether or not the glass furnace is of the regenerative type. The regenerative furnace requires periodic side-to-side switching of the fuel delivery. This can generally be accomplished with either a) side-to-side redundancy in the delivery system where only one side delivers petcoke at a given time while the other introduces purge air flow; or b) a single feed system with a switching mechanism coupled to recirculation lines that enable fuel and purge air to be diverted as needed for left or right side firing. While the redundant system (a) requires a higher initial capital cost, it is likely to afford smoother operation during transients, plus less piping complexity and simpler balancing of petcoke flow to each burner. The merits and drawbacks of both systems should be assessed for the particular glass melting application. In this regard, it should be mentioned that non-regenerative furnaces, such as occurs with oxy-fuel, offer the advantage of continuous petcoke delivery to the burners, which will lead to more stable combustion and process conditions than can be achieved in regenerator-based systems.

An additional key factor in material handling system design is whether or not dedicated feeders are used for each burner versus a single feeder with one or more discharge splitters for solid fuel delivery to two or more burners. This, however, requires that burners have been identified by furnace zone that can operate at nominally the same firing rate. Whether or not this level of precision and constraint are acceptable for glass furnace process control needs to be evaluated on a case-by-case basis. A practical example is in regenerative furnaces where fuel delivery to a given port is likely to come from a single feeder and is subsequently split to two or three fuel injectors, depending upon the port firing configuration. It is our experience that evenly splitting a solid fuel delivery header into 3 outlet streams is much more challenging than when only a two-way split is required. Either way, it is crucial to include flow balancing devices such as adjustable/removable orifices or riffles that can be "tuned" during operation to achieve a desired uniformity in flame appearance. We have also found oxygen enrichment at the injector level to be an effective tool in assisting to mitigate the effects of solid fuel flow imbalance on flame appearance and heat release. An example of this is provided in the Case Studies.

CASE STUDIES OF GLASS MELTING WITH OXYGEN-ENRICHED SOLID FUEL COMBUSTION

The following several case studies are intended to briefly highlight a cross-section of key results of our recent experience in oxy-solid fuel combustion for glass melting.

Case 1: Dark Specialty Glass Production using Combination of Oxy-Petcoke and Oxy-Natural Gas Firing

This application utilizes two SF oxy-petcoke burners at the batch end of the nominal 42 tpd cross-fired melter, while the remainder of the furnace comprises 8 Mini HRi™ oxy-gas flat flame burners. Individual feed systems driven by relatively low pressure transportation air blowers deliver the petcoke independently to the two solid fuel burners. The SF burners produce a stable, highly luminous and adjustable-length round solid fuel flame, with back-up gas or oil lances available for rapid fuel-switching between petcoke and other fossil fuels, as dictated for example by fuel price volatility and/or temporary interruption to the fuel supply. A photo of one of the SF oxy-petcoke flames in this furnace is presented in Figure 5. Note: The SF burners differ from the previously mentioned SF injectors in that the burners are designed for stoichiometric oxy-fuel combustion, while the injectors utilize only a small fraction of the oxygen required for combustion.

Figure 5. Photo of SF Oxy-Petcoke flame in glass melting furnace (Case 1).

Initial results of the switch from full oxy-gas to oxy-gas/petcoke have been favorable. Specifically, bottom temperature beneath the oxy/petcoke burners increased by 10 deg C, and glass quality improved. The fuel-switching experience was not, however, without its challenges, as there were several interruptions in the petcoke pneumatic delivery system early in the project caused by stray material in the fuel supply that required rapid and immediate removal of the SF burner's solid fuel nozzles and immediate replacement with backup oxy-natural gas lances. Overall, the customer is very satisfied with the operation, which has been ongoing for nearly two years, and plans to convert several additional burners to oxy-petcoke in the near future.

Case 2: Dark Specialty Glass Production using Combination of Oxy-Petcoke and Oxy-Fuel Oil Firing
Similar to Case 1, Case 2 is a dark specialty glass produced in a cross-fired oxy-fuel tank with 8 SF burners. Pull rate is nominally 50 tpd. Half of the burners are currently firing petcoke in the melting end of the tank, while the other half are firing heavy fuel oil in the fining end using s Gen1-SF backup oil nozzles. The customer reports that glass quality is as good now as with 100% oxy-oil operation, and plans to eventually convert to 100% oxy-petcoke.

Case 3: Clear Specialty Glass Production using Oxy-Petcoke/Coal Firing

This case involves a small, single-burner 15 tpd melting furnace previously fired with synthetic fuel gas. Notably, the furnace provides very little residence time to achieve complete combustion in comparison to a typical glass melting furnace. Conversion was initially to a single Cleanfire SF burner firing petcoke, but the customer was not satisfied with the color of the product (see Figure 6). Subsequently the customer switched from petcoke to coal and the color problem was resolved. Two factors were identified as being linked to the color problem; high vanadium content and poor combustion. Regarding vanadium content, Table IV summarizes key constituents found in the petcoke from Cases 1 and 3, as well as the coal from Case 3. Note that the magnitude of vanadium, a known coloring agent, in the petcoke from Case 3 was over 10 times higher than either the petcoke from Case 1 or the coal from Case 3.

Figure 6. Photo of glass product produced with syngas firing (left) and petcoke firing having high vanadium content and poor combustion (right).

Table IV
Values represent % of parent fuel

Element	Case 1: Petcoke	Case 3: Petcoke	Case 3: Coal
Vanadium	.018%	.190%	.012%
Iron	Not Sampled	.110%	.096%
Nickel	.0055%	.090%	.035%
Ash	.45%	1.19%	13.1%
Volatiles	10.7%	15%	29%

Combustion problems reported with the petcoke in Case 3 were traced back to a transportation air flow rate that was 2-3 times higher than recommended by Air Products. As such, ignition delay of the petcoke was unavoidable, resulting in significant unburned carbon. Hence, as previously suggested, it is clear that much of the ash-bound vanadium and carbon migrated to the glass melt where it led to redox changes and color contamination of the glass product. It is interesting that, despite the much higher ash content of the coal, the higher coal volatility nevertheless resulted in minimal ignition delay, good combustion and, hence, acceptable glass quality. While it was not possible to differentiate the relative effects on product color of vanadium content versus combustion quality, results from this case affirms the need for attention to be given to ash composition, and further underscores the importance of achieving good combustion.

Case 4: Clear Container Glass Production in a Regenerative Air-Fuel Furnace using Oxygen-Enriched Air Firing of Petcoke

This final case is a 200 tpd regenerative end port furnace producing clear container glass. The customer desired to replace heavy fuel oil with petcoke as the principal fuel, while oxy-natural gas boost was also supplied with Cleanfire Advanced Boost burners. We proposed using oxygen-enriched SF Port Injector technology instead of the combination of air-petcoke injectors plus oxygen lancing, as originally planned by the customer. The first step of this project was thus to compare the two approaches to petcoke combustion. Results are summarized in the photographs of Figure 7.

The photo on the left features three SF injectors, while that on the right shows three air-petcoke lances plus adjacent oxygen lancing. Firing rate and total oxygen enrichment is nominally the same for the two cases. The photos, taken with the same furnace camera, clearly demonstrate that the flames produced by the SF injectors were longer, broader and more luminous than those produced with air-fuel injectors plus O2 lancing. Moreover, close examination of these photos show that while three distinct flames are apparent for the SF injectors, only two can be discerned for the air-fuel lances. The issue here is that a single feeding system with splitting devices was used to supply all three under-port lances on each side of the furnace. However, the fuel split was plagued with imbalances which manifested itself principally in fuel deficiency to the outermost injectors. The fact that the outermost flame is visible, albeit relatively small, for the oxygen-enriched Port injectors, but not readily visible for the air-fuel injectors is, we believe, the result of the oxygen-fuel mixing facilitated by the Port injectors that helped to for the detrimental effects caused by the fuel imbalance. Based on these results the customer chose to move forward using the Port injectors. Note that no negative effects on glass quality have occurred, and while furnace

Figure 7. Photos of SF Port Injectors with integral oxygen mixing (left) and air-petcoke injector plus oxygen staging (right) in an end-port regenerative container glass furnace

generated NOx has increased somewhat, NOx emissions exiting the SCR system remained the same as they were with air-heavy oil firing. Two key takeaways from this case are that the controlled oxy-fuel mixing occurring within the Port injectors produces a superior flame to the combination of air-fuel injection plus oxygen lancing, and that even splitting of solid fuel streams in pneumatic conveying is inherently challenging and would certainly benefit from strategic placement of balancing devices, as previously suggested.

ECONOMIC CONSIDERATIONS

The economic viability of converting of a glass melting furnace from oil or gas to petcoke, as well as that of its ongoing operation, depends upon factors such as the cost of fuel, combustion system efficiency, glass quality and cost/frequency of a furnace rebuild. While not enough data exist to carry out a comprehensive cost study, we propose to compare differential operating costs from a base-case of air-natural gas firing in a regenerative furnace to petcoke firing in either another regenerative air-fuel system (Option A), or employing 100% oxy-petcoke operation (Option B). Key assumptions for the analysis are that:
1. Natural gas price is $18/MMBtu

2. Petcoke price is $9/MMBtu (pulverized and delivered)
3. Oxygen price is $60/MT
4. Pull rate of glass is 300 MTPD
5. Switching from natural gas to petcoke involves no change in fuel efficiency
6. Switching from air/fuel to oxy/fuel increases fuel efficiency by 20%

Results of the analysis, which reduce to a comparison of annualized fuel and oxygen costs, are summarized in Figure 8, adapted from Goruney et al [8]. Note that both petcoke Options A (air-fuel) and B (oxy-fuel) lead to a substantial reduction in fuel cost from the base air-natural gas firing case; from an air-natural gas baseline of $13MM/yr to $6.4MM/yr for Option A and $5MM/yr for Option B. However, the addition of oxygen in Option B adds $2.8MM/yr of oxygen cost, resulting in an apparent net operating cost increase relative to Option A of approximately $1.4MM/yr, or approximately $12.80/MT of pulled glass. The question then becomes whether or not the higher operating cost with oxygen is offset by benefits related to lower capital cost, longer furnace campaigns and superior glass quality control due improved combustion efficiency and stability. Regarding capital costs, a recent study suggests that, relative to a regenerative air-fuel furnace, oxy-fuel capital costs are lower by 30-40% [9]. Moreover, our discussions with glass manufacturers having experience with air-fuel petcoke combustion confirms combustion quality problems and suggests that furnace rebuilds as frequently as every 2 to 3 years is not uncommon. When we contrast this with the nearly two year successful run as documented herein in Case 1, we believe

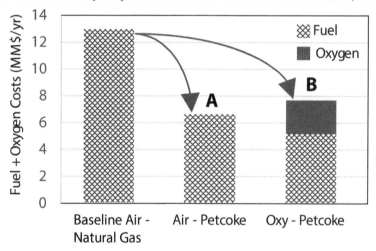

Figure 8. Graph of fuel plus oxygen cost for 300 MTPD glass melting operation associated with different firing scenarios.

there is a very persuasive case to be made that oxygen is a vital ingredient in the mix of factors needed for cost-effective glass melting with petcoke.

CONCLUSION

Glass manufacturers have turned to petcoke and other relatively low-cost solid fuels in order to maintain acceptable financial performance during this period of regional fuel price volatility and reductions in production demand. Through experience gained in this area as both an oxygen and combustion technology supplier, we have summarized herein several key findings that are essential to a successful adaptation of solid fuel firing for glass melting. Prior to making the fuel switch, attention must be given to compositional impurities in the fuel, and how they could potentially effect both glass product quality and furnace life. The most important factor during operation is the attainment of complete combustion, which has implications extending beyond melting efficiency, and into the realm of glass redox, color, defects and furnace life. Two essential elements needed to achieve a consistently high-performance solid fuel combustion system are a robust, well-designed material handling system, and oxygen enrichment for combustion. To these points, we have illustrated how the material handling system must be well-matched with the combustion enabling equipment, and that optimization of the oxygen-fuel mixing processes is needed to ensure that the full benefit of oxygen can be realized. Because, even in the best case, solid fuel material handling systems are less reliable than those for oil and gas, on-the-fly back-up fuel firing capability is needed. Finally, serious consideration should be given to full oxy-solid fuel firing, not only for combustion and glass quality reasons, but also to eliminate the need for regenerators which have the potential to be a focal point for fouling and corrosion during solid fuel operation, and are likely to be the largest single factor leading to reduced furnace performance and shortened furnace life.

REFERENCES
1.   Man, C. K., and Gibbins, J. R., "Factors Affecting Coal Particle Ignition Under Oxyfuel Combustion Atmospheres", Fuel, Volume 90, Issue 1, pp. 294-394, January, 2011.
2.   "Spontaneous Breakage of Fully Tempered Glass", Eckelt Glas, GmbH, Edition 01.2003 (2005).
3.   Ito, T., et al, "Long-Term Stable Operation of Petroleum-Coke-Fired Boiler Retrofitted from Existing Oil Firing", Proceedings of the Engineering Foundation Conference on Inorganic Transformation and Ash Deposition During Combustion, pp. 895-914, 1991.
4.   Walsh, P. M., Mormile, D. J., and Piper, B. F., "Sulfur Trioxide Formation in the Presence of Residual Oil Ash Deposits in an Electric Utility Boiler", pp. 294-300.
5.   "Integrated Pollution Convention and Control", Reference Document on Best Available Techniques in the Glass Industry, Prepared by the European Commission, December 2001
6.   Lewandowski, D., "Design of Thermal Oxidation Systems for Volatile Organic Compounds", CRC Press, 2000.
7.   IMI-NFG Course on Processing in Glass, Lecture 2, Melting and Fining Processes in Industrial Glass Furnaces, Celsian Glass and Solar, presented at Lehigh University, Spring 2015.
8.   Goruney, T., et al., "Alternative Fuels for Glass Melting", Glass International, October 2013.
9.   Worrell, E, et al., "Energy Efficiency Improvement and Cost Saving Opportunities for the Glass Industry", an ENERGY STAR® Guide for Energy and Plant Managers, sponsored by the US EPA, March 2008.

HYDROPRIME® GAS GENERATORS PROVIDE LOW COST, RELIABLE HYDROGEN FOR FLOAT GLASS MANUFACTURING

Goutam Shahani, Kyle Finley, Nick Onelli
Linde Group
Blue Bell, Pennsylvania, USA

ABSTRACT

In the manufacture of float gas, hydrogen and nitrogen are used in the tin bath to provide the appropriate reducing atmosphere. This paper provides an overview of Hydroprime®, a unique steam methane reformer, employing an integrated heat recovery system, which reduces the cost and improves the reliability of hydrogen supply. Hydroprime plants can produce 0.15 – 0.3 MMSCFD of 99.999 % hydrogen at 200 psig. This system offers a flexible hydrogen supply solution for the float glass industry, which has demonstrated excellent results in actual commercial operation worldwide.

INTRODUCTION

Hydrogen and nitrogen atmospheres are used as a mixture in a float glass line to prevent the tin bath from getting oxidized to tin oxide (SnO_2). The formation of SnO_2 will cause flaws in the glass surface. The typical hydrogen concentration for this process is 5 – 10%. Relatively small volumes of hydrogen <1,095 Nm^3/h (1 MMSCFD) are supplied in bulk by trailer or produced on-site by electrolysis, methanol and ammonia dissociation or steam methane reforming. While steam methane reforming is the dominant method of producing hydrogen at a relatively large scale >1,095 Nm^3/h (1 MMSCFD), this technology has not yet been widely adopted at a small scale due to cost and reliability considerations.

This paper will provide an overview of an innovative, hydrogen generator based on proven steam methane reforming technology. Linde's HYDROPRIME gas generators are compact, efficient, and flexible. Ten units have been built, of which nine have been placed in commercial operation for an extended period of time, thereby demonstrating performance in a variety of applications all over the world. A picture of one such gas generator is presented below in Figure 1.

Figure 1. HYDROPRIME Gas Generator

HYDROPRIME GAS GENERATOR

Technology

The predominant method of producing hydrogen on an industrial scale is steam methane reforming. The gas generators employ a unique heat integration concept, while using this proven technology. In this process, desulfurized natural gas feed is mixed with preheated water and fed to tubes filled with nickel catalyst. The following reactions take place at elevated temperature and pressure:

Reforming

$$CH_4 \quad + \quad H_2O \quad \leftrightarrow \quad CO \quad + \quad 3H_2 \quad \text{Endothermic} \quad (1)$$

Shift Reaction

$$CO \quad + \quad H_2O \quad \leftrightarrow \quad CO_2 \quad + \quad H_2 \quad \text{Exothermic} \quad (2)$$

Approximately 75% of the conversion to hydrogen takes place in reaction (1). Reaction (2) drives the equilibrium balance further to yield a hydrogen rich gas. Reaction (1) is reforming which is endothermic; reaction (2) is shift conversion which is exothermic. Both reactions (1) and (2) take place in the reformer. However, only reaction (2) takes place in the shift converter and employs promoted iron oxide catalyst. Both reactions are equilibrium limited, based on the outlet temperature and pressure. The reaction products are a mixture of H_2, CO,

CO_2 and H_2O. A simplified block flow diagram of the SMR process is presented below in Figure 2.

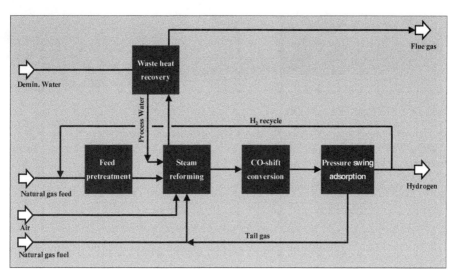

Figure 2. Steam Methane Reforming (SMR)

The overall reforming reactions are endothermic, requiring heat, which is supplied by the combustion of fuel. The tail gas from the PSA system meets most of the fuel requirement with the rest being supplied by natural gas.

The hydrogen rich stream is purified by pressure swing adsorption (PSA). The PSA is a physical process that depends on the selective, physical binding of gas molecules. Hydrogen, being non-polar and highly volatile, is essentially not adsorbed by the proprietary adsorbent material, which is a mixture of carbon molecular sieve and zeolites. The system operates on a repeated cycle having two basic steps: adsorption and regeneration. The regeneration consists of de-pressuring, purging and re-pressuring.

Design

The inputs to this modular gas generator are natural gas, demineralized water, and electric power. The outputs are high purity, gaseous H_2. There is no export steam. This gas generator produces H_2 with the following specifications.

- Flow rate from 165 – 330 Nm^3/h (0.15 - 0.3 MMSCFD)
- Purity of 99.999 +%
- Pressure of 13.8 barg (200 psig)

A picture of a partly assembled gas generator in transit is presented below in Figure 4.

Figure 3. Gas Generator in transit

The major dimensions of the skid are listed below in Table I. The gas generator is designed to be easily shipped all over the world. It can be placed in a flat-bed truck, thereby simplifying logistics and transportation.

Table I. Gas Generator Skid Dimensions

Dimension	Module, m
Length	14
Width	3
Height	4

These gas generators are heat integrated for high thermal efficiency. The typical operating characteristic for a 330 Nm³/hr (0.3 MMSCFD) gas generator are summarized below in Table II. These results have been achieved in actual commercial operation over an extended period of time.

Table II. Gas Generator Performance

H_2 capacity, Nm³/h	330
H_2 purity, %	99.999
$CO+CO_2$, ppm	< 1
Pressure, barg	13.8
Natural gas /H_2 MM cal/Nm³ H_2 (HHV)	4.4
Power, KW	60
Demineralized water, kg/h	550

The gas generators have a modular, open skid design which provides the following advantages:

- Simple and quick site installation
- Outdoor installation
- Excellent accessibility for maintenance
- Small foot print

These gas generators are standardized with fully automatic fail-safe controls, allowing unattended operation with remote start-up and monitoring 24/7. This enables quick response to even the slightest production interruption. Furthermore, the system provides virtually 100% uptime for H_2 product supply with the addition of a simple liquid backup tank.

BENEFITS

The gas generators offer a very viable alternative for relatively small <1,095 Nm^3/h (1 MMSCFD) hydrogen requirements. The gas generators are shop fabricated with high labor efficiency in a controlled environment, ensuring high quality control. The modular design of these gas generators reduces field construction, minimizes risk and provides faster project schedule. These gas generators also provide superior environmental and safety performance based on low emissions and reduced truck deliveries. Furthermore, the combination of low utility consumption and high reliability offers a very cost competitive solution for relatively small, industrial hydrogen consumers. Linde can provide 'over-the-fence' supply of H_2 molecules. In this arrangement, the end-user purchases the molecule, under long term contract eliminating the need for any capital investment and responsibilities related to ownership such as installation, operation, maintenance, repair, insurances, manning and back-up of product in case of plant stoppage. Alternatively, these gas generators can also be owned and operated by the end-user.

CONCLUSIONS

A HYDROPRIME gas generator is a unique standardized steam methane reformer developed to reduce the cost and improve the reliability for relatively small industrial hydrogen consumers. These gas generators offer many advantages over traditional supply modes such as electrolytic plants, conventional steam methane plants, and truck-delivered bulk hydrogen. The gas generators are highly heat integrated, which translates into low operating cost. Furthermore, these gas generators offer high reliability, superior environmental and improved safety performance. These modular hydrogen gas generators are fully automatic with fail-safe controls, allowing unattended operation with remote start-up and monitoring. In addition, these gas generators are modular with an open skid design, which provides easy site installation and excellent accessibility for maintenance. These gas generators provide a very competitive cost of ownership. The gas generator offers a flexible hydrogen supply solution for the float glass industry. These gas generators have demonstrated excellent results in actual commercial operation throughout the world.

CORROSION OF AZS REFRACTORIES – SOURCE OF DEFECTS IN TABLEWARE GLASS

Šimurka P.[1], Kraxner J.[1], Vrábel P.[2], Paučo T.[2]
[1] *Vitrum Laugaricio, Joint Glass Centre of the II SAS, TnU AD and FCHFT STU, Trenčín, 911 50 Slovakia*
[2] *Rona a.s., 020 61 Lednické Rovne, Slovakia*

ABSTRACT
    The relationship between the composition of defects present at tableware glass products and a glass phase of the Alumina Zirconia Silicate (AZS) refractory material is studied. The refractory samples have been taken from the working and melting parts of a furnace after production termination as well as from the material where the laboratory static corrosion tests have been conducted. The concentration profiles of oxides, present in the AZS refractory glass phase, were determined by scanning electron microscopy (SEM) and energy dispersive x-ray (EDX) analysis. Samples of the final products containing different types of inhomogeneities were collected during regular production in a period of 48 months and analyzed. To identify potential sources of defects, the results of this refractory corrosion study were compared with the composition of defects present in the final products of tableware glass production.

INTRODUCTION
    A fused cast Alumina Zirconia Silicate (AZS) is commonly used type of refractory material in glass industry because of its remarkable resistance to corrosion by glass melt and glass melting environment. The superior corrosion resistance of AZS refractories is primarily attributed to their unique microstructural features. The fused cast AZS refractories posses almost no porosity, which is one reason for the high corrosion resistance to glass melts. Mineralogically, bodies of fusion-cast AZS refractory used by industry have been a mixture of ZrO_2, Al_2O_3, and glassy phase that contains mostly SiO_2. The crystals of the main phase (zirconia, alumina) have an intercrystalline bonding and the glassy phase only exists in the cavities between the crystals without bonding feature. The glassy phase of AZS refractories is important for its production process. The fused cast materials are molten by means of arc welding in electric melting aggregates. The main components of the glassy phase (SiO_2, alkalis) lower the melting point of the AZS materials thus enabling a simpler and less expensive production. Moreover, on cooling the fused cast blocks the glassy phase decreases the occurrence of edge cracks. The glassy phase can relax or reduce mechanical stress present in the blocks during the heating blocks at high temperature. The thermal shock resistance of the fused cast materials is also improved by the glassy phase as far as necessary for the application in glass melting tanks. [1-4]
    Despite the above advantages, the presence of the glassy phase is realized to be a problem from the viewpoint of applications. While there can be many sources of cord defects, refractory contamination in glass is frequently cited as a leading source. Glass contamination can and does occur from the exudation and corrosion of superstructure and glass contact fusion-cast AZS refractories. [5]
    The glassy phase exudation from fused cast AZS can occur at high temperatures due to the expulsion and the runoff of a part of the glassy phase at the surface of the refractory. The behavior of AZS during heating up but also during the campaign can give rise to problems directly affecting glass quality by generating defects in glass such as vitreous enriched alumina defects with or without secondary zirconia crystals.

The corrosion of refractory materials causes the formation of a boundary layer that is enriched in the components of the refractory material. When this boundary layer ends up in the bulk of the melt, it forms a cord. [6-8]

A cord is a vitreous particle with a composition deviating from the composition of the bulk of the glass, and consequently, a different index of refraction, which brings about a local lens effect. The vitreous particle may contain crystals. [9]

To eliminate a cord defect problem in the practice of industrial glass production, an indication of the origin of the cord in the furnace is very helpful.

Objective of this paper is to show the relationship between the chemical composition of the cords in the glass products and their origin. By combining information on the chemical composition of the cords with data on the chemical composition of the glass melt/AZS interface obtained from the AZS refractory glassy phase analysis, the possibility to link a production cord defect to its origin area, where it has been formed in the refractory, is shown.

EXPERIMENTAL

The samples of corroded refractory material AZS32 were taken from the sidewalls of the melting and working parts of glass furnace near the throat. The recuperative furnace producing crystalline glass has been in operation for 89 months, melting temperature in the furnace was in the range of 1470–1480 °C, temperatures in the working part were from 1280 to 1340°C and daily pull was between 13-30 tons of glass. The static corrosion test was performed in an electric furnace at 1475°C for 96 hours, in a platinum crucible filled with the crystalline glass containing 5.2 % K_2O, 11.8 % Na_2O, 5.0 % BaO, 8.0 % CaO, 1.0 % Al_2O_3 and 68.1 % SiO_2. The new refractory sample AZS 32 105 × 25 ×15 mm parallelepiped shaped has been used. During the test, the sample was kept immersed inside the glass for about 40 mm. The samples were extracted from the hot glass after 96 hours. Static corrosion test was realized in SSV Murano, Italy. The samples of refractory materials were cut off from the corroded refractory stone, subsequently polished and finally cleaned in ultrasound bath. The surface of the sample was sputtered by carbon and examined by electron microscopy. The elemental composition and concentration profile of individual oxides, starting at glass-refractory boundary and continuing into refractory material, was determined by scanning electron microscopy/energy-dispersive x-ray (SEM/EDX) spectrometer (JEOL JSM 7600F/EDS/WDS/EBSD). The samples of the final products containing different types of cords were collected during regular production in a period of 48 months. The composition of defects was analyzed by SEM/EDX analysis in Glass Service a.s., Vsetín, Czech Republic.

THE DATABASE OF CHEMICAL COMPOSITIONS OF CORDS PRESENT IN THE FINAL PRODUCTS

Database of 69 chemical compositions of defects, mainly cords, have been created by collecting of the final products containing glass defects during regular production. These products were collected in the periods when high rejection due to the defect presence in daily production was observed. The products were coming from different furnaces in which the glass melt with the same glass composition was prepared. The range of composition of the individual oxides in glass defects together with the composition of glass surrounded defects (Glass) and composition of the glassy phase in the non-used AZS refractory material (AZS) in wt. % are in Table 1. The composition of surrounded glass was calculated as the average value from analysis of 30 samples, the composition of the glassy phase as the average value from the analysis in 4 different locations of non-used refractory material.

Table 1. Range of compositions of the individual oxides in glass defects, composition of glass surrounded defects and glassy phase in non-used AZS refractory material in wt.%

Oxid	Na$_2$O	Al$_2$O$_3$	SiO$_2$	K$_2$O	CaO	ZrO$_2$	BaO
Range (wt. %)	7.6-12.8	0.8-32.2	47.4-75.4	5.2-10.0	0.3-8.3	0.0-6.4	0.4-4.7
Glass (wt. %)	10.6	1.2	71.5	5.3	7.1	0.0	4.4
AZS (wt. %)	2.6	17.6	77.0	0.0	0.5	1.1	0.0

The content of the individual oxides present in the cords sorted from the lowest to the highest value is shown in the Figures 1 – 7.

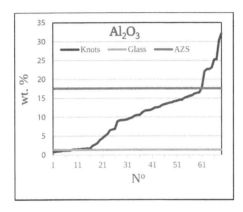

Figure 1. Content of Al$_2$O$_3$ in the cords sorted from the lowest to the highest value

In Figure 1 we can see that there are 16 defects, where the content of Al$_2$O$_3$ is close to its content in the original glass (difference is not higher than 0.5 wt. %). 9 defects have higher content that is average value of Al$_2$O$_3$ in the glassy phase of non-used refractory. Figure 2 shows that no defect reaches the content of SiO$_2$ present in the glassy phase of AZS refractory and 55 defects have content of SiO$_2$ lower than it is in the original glass.

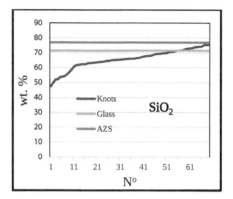

Figure 2. Content of SiO_2 in the cords sorted from the lowest to the highest value

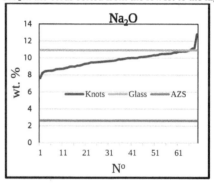

Figure 3. Content of Na_2O in the cords sorted from the lowest to the highest value

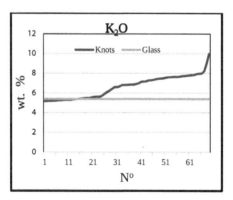

Figure 4. Content of K_2O in the cords sorted from the lowest to the highest value

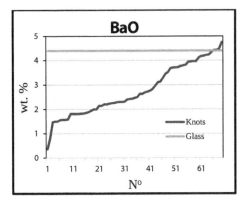

Figure 5. Content of BaO in the cords sorted from the lowest to the highest value

It can be seen in Figure 3, that besides 3 defects, all other ones have content of Na_2O like in the original glass or less. The range of content is practically the same like in K_2O. While content of Na_2O is lower than in the original glass, the content of K_2O in defects is like in the original glass (about 20 defects) or higher (Figure 4).

Figure 5 shows that almost all defects have lower content of BaO than it is in the original glass and the range of content BaO in the cords varies from 0.4 to 4.7 wt. %.

About 60 defects has higher content of CaO in comparison with its content in the glassy phase of refractory and simultaneously lower content then it is in the original glass. Only in 9 cords the content of CaO exceeds this value (Figure 6).

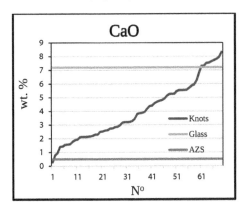

Figure 6. Content of CaO in the cords sorted from the lowest to the highest value

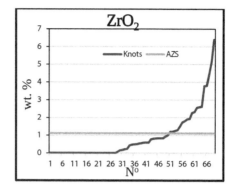

Figure 7. Content of ZrO_2 in the cords sorted from the lowest to the highest value

In Figure 7, there are 28 defects without content of ZrO_2. The rest of defects is divided proportionally - one part has content of ZrO_2 like in the glassy phase of refractory or lower, the second part of defects has higher content of ZrO_2. The range of ZrO_2 content in cords varies from 0 to 6.4 wt. %.

Both, superstructure corrosion and corrosion of glass contact AZS refractory, results in the formation of cords in which Al_2O_3 content is higher in comparison with the content in the original glass melt [9]. Therefore the content of Al_2O_3 can be the one criteria for the identification of the cord source and it´s linking to the origin area, where it had been formed. In Table 2, there are chemical compositions of the cords, where the difference of content of Al_2O_3 is not higher than 0.5 wt.% in comparison with its content in the original glass. Simultaneously the content of other oxides is not far from their content in the original glass. It is possible to say that origin of these defects is not in the AZS material. Possible reasons of their creations can be wrong batch weighing and/or mixing, wrong granulometry of raw materials, using of not suitable cullet, evaporation of alkali from the glass melt surface and so on.

Table 2. Chemical composition of cords with content of $Al_2O_3 < 2.0\%$ (wt.%)

	Na_2O	Al_2O_3	SiO_2	K_2O	CaO	ZrO_2	BaO
1	10.51	1.70	73.91	5.20	5.53	0.00	3.77
2	10.12	1.70	74.35	5.23	5.26	0.00	3.96
3	10.64	1.68	72.98	5.22	5.53	0.00	3.95
4	9.10	1.59	71.62	5.47	8.30	0.21	3.72
5	10.76	1.57	69.86	5.27	7.94	0.17	4.42
6	10.10	1.50	75.33	5.18	4.81	0.00	3.52
7	10.54	1.47	70.73	5.50	7.80	0.00	4.42
8	10.22	1.45	73.10	5.32	5.69	0.00	4.22
9	10.17	1.21	71.87	5.20	7.53	0.00	4.10
10	10.26	1.17	71.43	5.16	7.63	0.00	4.34
11	9.73	1.13	74.00	5.52	5.47	0.00	4.17
12	10.00	1.20	74.27	5.39	5.55	0.00	3.79
13	9.52	1.00	75.30	5.26	5.25	0.00	3.68
14	9.29	0.97	75.44	5.37	5.12	0.00	3.82
15	10.32	0.85	71.59	5.30	7.72	0.00	4.23
16	9.18	0.79	73.38	5.43	7.30	0.00	4.74
Glass	10.6	1.2	71.5	5.3	7.1	0.0	4.4

In Table 3, there are chemical compositions of cords with content of Al_2O_3 higher than 5.0 wt. %. The content of K_2O is higher and the content of CaO and BaO is lower in comparison with their content in the bulk glass, while the content of Na_2O varies from 8.24 to 12.82 wt. %. There are 11 cords without the presence of ZrO_2. In Table 4, there are the couples of cords with the comparable chemical composition; they differ mainly by the presence of ZrO_2. As the content of all other main oxides of these couples is very close each to other, we can say that these cords originate from the same kind of source. A cord coming from the AZS refractory glassy phase may not contain ZrO_2; the content of other oxides, mainly K_2O, plays important role at determination of the cord origin.

Table 3. Chemical composition of cords with content of $Al_2O_3 > 5.0\%$ (wt. %)

	Na_2O	Al_2O_3	SiO_2	K_2O	CaO	ZrO_2	BaO
1	12.82	32.15	47.44	5.62	0.78	0.83	0.36
2	8.47	30.3	49.90	9.98	0.29	0.55	0.79
3	10.75	25.33	52.10	7.99	0.96	1.8	1.82
4	11.20	25.2	51.93	7.73	1.4	1.50	1.59
5	10.78	23.11	54.32	7.59	1.55	0.83	1.81
6	10.77	22.67	53.80	7.64	1.60	1.73	1.81
7	9.57	22.66	55.41	7.96	2.12	0.59	1.81
8	10.83	22.08	53.59	7.18	2.86	1.00	2.00
9	10.48	17.70	61.77	7.14	1.43		1.47
10	8.70	16.56	63.00	8.23	1.94		1.57
11	8.73	16.39	63.13	7.85	2.5		1.85
12	9.66	16.19	62.29	7.55	2.29		2.3
13	8.49	15.74	64.05	7.76	2.16		1.81
14	10.00	15.60	58.67	5.48	6.24	1.29	2.72
15	8.68	15.32	64.92	7.67	1.78	0.15	1.50
16	9.37	15.1	63.37	7.63	1.57	1.2	2.5
17	8.48	14.59	65.50	7.80	2.11	0.07	1.57
18	9.47	14.56	63.07	7.61	1.90	2.32	1.5
19	8.99	14.31	62.29	6.60	4.41	1.25	2.14
20	9.50	14.23	64.46	7.48	2.14		2.20
21	9.64	13.91	61.66	7.47	3.88	1.2	2.43
22	8.40	13.80	65.42	7.94	2.61		1.83
23	10.32	13.61	56.96	8.97	3.28	3.80	3.7
24	9.61	13.41	63.71	7.72	2.83	0.77	1.94
25	9.1	13.19	66.00	7.64	2.2		2.14
26	9.54	12.73	65.31	7.4	2.12	0.82	2.47
27	9.68	12.57	60.73	6.86	2.73	5.15	2.29
28	8.24	12.54	65.89	7.81	3.21		2.30
29	9.88	12.13	64.49	6.99	2.59	1.92	2.00
30	9.58	11.86	63.92	7.45	2.74	2.56	1.89
31	8.99	11.8	67.47	6.80	3.20	0.23	2.25
32	8.88	11.67	65.70	7.16	3.85	0.43	2.31
33	9.93	11.26	66.04	6.82	3.16	0.53	2.26
34	10.0	10.6	65.8	6.6	2.3	2.6	2.2
35	9.48	10.6	66.53	6.915	2.52	2.27	2.23
36	8.97	10.4	63.54	6.86	3.46	4.44	2.69
37	8.55	9.93	67.84	7.33	3.95		2.4
38	8.67	9.74	67.09	6.81	4.38	0.47	2.83
39	10.5	9.42	66.42	6.84	3.8	2.62	1.56
40	9.99	9.32	62.18	6.12	3.20	6.4	3.13
41	9.99	9.23	65.11	6.44	5.54	0.59	3.10
42	9.32	9.19	67.77	7.3	2.97	1.9	2.63
43	8.8	8.88	68.02	7.28	4.19	0.80	2.76
44	9.83	6.89	69.27	6.67	4.92		2.42
45	9.84	6.73	69.03	6.26	4.59	0.93	2.63
46	10.48	6.55	65.11	5.64	4.69	3.78	3.71
47	10.64	5.50	68.47	5.77	5.85	0.49	3.28

Table 4. Comparison of the chemical composition of cords with and without ZrO_2 (wt. %)

	Na₂O	Al₂O₃	SiO₂	K₂O	CaO	ZrO₂	BaO
12	9.66	16.19	62.29	7.55	2.29		2.3
16	9.37	15.1	63.37	7.63	1.57	1.2	2.5
13	8.49	15.74	64.05	7.76	2.16		1.81
15	8.68	15.32	64.92	7.67	1.78	0.15	1.50
18	9.47	14.56	63.07	7.61	1.90	2.32	1.5
20	9.50	14.23	64.46	7.48	2.14		2.20
22	8.40	13.80	65.42	7.94	2.61		1.83
24	9.61	13.41	63.71	7.72	2.83	0.77	1.94
25	9.1	13.19	66.00	7.64	2.2		2.14
26	9.54	12.73	65.31	7.4	2.12	0.82	2.47
27	9.68	12.57	60.73	6.86	2.73	5.15	2.29
28	8.24	12.54	65.89	7.81	3.21		2.30
38	8.67	9.74	67.09	6.81	4.38	0.47	2.83
44	9.83	6.89	69.27	6.67	4.92		2.42
45	9.84	6.73	69.03	6.26	4.59	0.93	2.63

THE STUDY OF CONCENTRATION PROFILES OF THE AZS REFRACTORY GLASS PHASE

A microstructure of non-used AZS material is illustrated in Figure 8. The chemical composition of the glass phase from this AZS material analyzed by EDX in 4 different locations is in Table 5.

Figure 8. Microstructure of the AZS refractory of a new AZS refractory material

New fused cast AZS 32 refractory material consists of about 48 % corundum, 32 % baddeleyite and 20 % of glassy phase (wt. %) [11]. The glass phase of refractory contains about 1.5-4.0 % Na₂O, 16.0 – 19.5 % Al₂O₃, 75.0-79.5 % SiO₂, 0.5-1.5 % ZrO₂, less than 1.0 % CaO, TiO₂ and Fe₂O₃ (in wt. %) (Table 5).

Table 5. Chemical composition of glass phase in a new AZS 32 in different locations (wt. %).

	Na₂O	Al₂O₃	SiO₂	CaO	ZrO₂	TiO₂	Fe₂O₃
1	2.59	18.57	76.38	0.55	0.84	0.24	0.83
2	4.00	16.51	76.92	0.39	1.26	0.28	0.64
3	2.27	19.13	75.35	0.68	1.14	0.41	1.00
4	1.65	16.31	79.35	0.50	1.18	0.28	0.74

Microstructure of the AZS refractory material taken from the melting part of the furnace after campaign located above the metal line consists only from the ZrO_2 (white) and glass phase (grey) (Figure 9). Al_2O_3 has been completely dissolved in the glass phase.

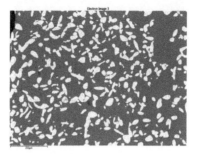

Figure 9. Microstructure of the AZS refractory located in the melting part above the metal line

The concentration profile of the AZS refractory glass phase taken from the melting part of the furnace after campaign analyzed by EDX is in Table 6. The glass phase of refractory located above the metal line is enriched about BaO and K_2O in comparison with their content in a new AZS refractory material. The content of individual oxides, starting at glass-refractory boundary and continuing into the refractory material has no tendency of increasing or decreasing. Besides K_2O, content of alkali oxides in refractory glass phase is lower in comparison with their content in the original glass melt. The content of Al_2O_3 in the glass phase is about 30 wt. %. The content of ZrO_2 is practically the same like in a new AZS refractory.

Table 6. Chemical composition of glass phase in AZS refractory located in the melting part above the metal line (wt. %).

d(mm)	Na_2O	Al_2O_3	SiO_2	K_2O	CaO	ZrO_2	BaO
0.04	8.25	30.64	47.34	6.26	2.43	1.01	4.07
1.01	8.35	30.51	46.58	5.90	2.52	1.86	4.27
3.27	8.26	32.38	44.76	5.74	3.00	1.01	4.83
5.09	8.46	29.61	46.85	5.87	2.99	0.70	5.09
10.04	8.77	29.96	47.00	6.04	2.81	0.67	4.75

Microstructure of the AZS refractory material taken from the melting part of the furnace after campaign located below the metal line is illustrated in Figure 10; microstructure of the AZS refractory material taken from the working part of the furnace after campaign located below the metal line is illustrated in Figure 11.

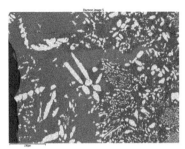

Figure 10. Microstructure of the AZS refractory located in the melting part below the metal line

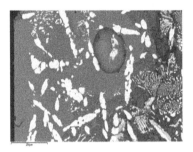

Figure 11. Microstructure of the AZS refractory located in the working part below the metal line

Microstructure of the AZS refractory, analyzed after the laboratory static corrosion test, is in Figure 12. The sample for the analysis was taken from the part below the metal line.

Four phases (ZrO_2, Al_2O_3, glass phase and a new phase) can be observed in all AZS refractory materials located below the metal line. They differ by presence and by the width of the boundary layer of glass phase (Figures 10-12).The composition of a new phase is close to a mineral nepheline ($(Na,K)AlSiO_4$).

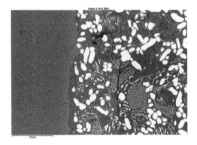

Figure 12. Microstructure of the AZS refractory analyzed after laboratory static corrosion, located below the metal line

Table 7. Chemical composition of glass phase in AZS refractory from the melting part located below the metal line (wt. %).

d (mm)	Na₂O	Al₂O₃	SiO₂	K₂O	CaO	ZrO2	BaO
0.02	8.70	22.68	52.41	8.44	2.14	2.20	3.42
0.21	8.90	23.24	51.04	7.89	2.25	2.39	4.28
0.38	8.75	24.58	49.86	7.82	2.33	1.56	5.11
0.56	8.16	23.82	50.70	7.79	2.48	1.42	5.63

Table 8. Chemical composition of glass phase in AZS refractory from the working part located below the metal line (wt. %).

d (mm)	Na₂O	Al₂O₃	SiO₂	K₂O	CaO	ZrO2	BaO
0.03	7.77	10.73	55.90	6.18	4.65	10.06	4.69
0.11	7.87	11.56	56.03	6.02	4.68	9.10	4.74
0.23	8.01	17.04	56.31	7.15	2.97	4.43	4.10
0.48	8.65	23.29	50.49	7.33	2.57	2.08	5.58

The concentration profiles of the AZS refractory glass phase taken from the melting and working parts located below the metal line are in Tables 7-8. As it can be seen in Tables 6-8, content of K₂O in the glass phase of AZS refractories taken after furnace campaign is higher in comparison with its content in the bulk glass. This up-hill diffusion occurs because in fact, not the gradient in the concentration profile but the gradient in the chemical potential respectively the activities of the diffusing cations determine the direction of the diffusion. If the chemical potential of the potassium oxide is lower in the AZS glassy phase, the system will try to shift into the direction of the lowest Gibbs energy. The cations will therefore diffuse into the glassy phase of the refractory until at last the activities are equal. If the chemical composition of the glass, respectively the glassy phase of the fused cast AZS in the transition layer is studied, one sees that the chemical composition does not follow the direct route from the glass composition to the composition of the glassy phase of the original fused cast AZS. Dietzel [12] explains this behavior, by assuming that the glassy phase tries to adopt a structure that resembles that of a (thermodynamic relatively stable) crystal in the neighborhood of the concentration route. The crystal structure distinguishes itself by its large thermodynamic stability and low Gibbs energy. Examples are: crystal formation of sodium nepheline (Na₂O.Al₂O₃.2SiO₂), leucite (K₂O.Al₂O₃.4SiO₂), or kaliophelite/kalsilite (K₂O.Al₂O₃.2SiO₂). The „ unusual" behaviour of the K₂O concentration profile at the interface and deeper in the AZS might therefore be explained by attempts to adopt the concentration of these the termodynamically stable crystals. [9].

The concentration profile of the AZS refractory glass phase after performing of the laboratory static corrosion test is in Table 9.

Table 9. Chemical composition of glass phase in AZS refractory after laboratory static corrosion test located below the metal line (wt. %).

d(mm)	Na₂O	Al₂O₃	SiO₂	K₂O	CaO	ZrO2	BaO
0.02	10.15	3.80	68.76	3.97	7.68	0.85	4.80
0.11	9.32	10.25	66.21	5.34	4.16	1.30	3.42
0.18	9.46	14.77	62.18	6.29	2.33	2.36	2.61
0.62	10.36	22.46	53.65	6.58	1.64	2.73	2.58
0.91	10.33	26.70	46.53	5.60	2.38	3.03	5.43

The glass phase below the metal line is enriched about BaO and K₂O in comparison with a new AZS refractory material. The concentration profile of individual oxides, starting at glass-refractory boundary and continuing into the refractory material, shows that, while the content of Al_2O_3, K_2O and ZrO_2 increases, the content of SiO_2, CaO and BaO decreases with a distance from the refractory-glass melt boundary. The content of Na_2O varies between 9.3 to 10.4 wt. % with no tendency of increasing or decreasing.

THE IDENTIFICATION OF POTENTIAL SOURCES OF CORDS

Glass contamination can and does occur from the exudation and corrosion of superstructure and glass contact fusion-cast AZS refractories. There is a difference between AZS exudation and corrosion of superstructure refractory in industrial glass-melting operations [5].

AZS exudation of superstructure may be characterized by:

- Occurrence due to heating in the presence of air, in the absence of corrosive species such as NaOH, KOH, batch dusts, etc.
- A viscous aluminosilicate phase coating the AZS surface and containing a minor quantity of zircoia crystals, which rapidly separate by gravity. The aluminosilicate cools to a transparent glass.
- In the absence of temperature cycling, exudation is a one-time rapidly exercised event that cannot continue perpetually.

AZS corrosion of superstructure may be characterized by:

- Occurrence due to heating in the presence of corrosive species (which dissolve crystalline alumina).
- A viscous aluminosilicate phase coating the AZS refractory surface, which upon cooling becomes an opaque glassy phase containing zirconia and alumina crystals.
- With or without cycling of environmental conditions, this phenomenon continues perpetually throughout the refractories´service life´, and can potentially cause ongoing defect generation in the glass manufacturer´s product .

The corrosion mechanism involves indiffusion of alkali species (both vapor phase and batch dust particles) from the furnace atmosphere, which causes dissolution of the AZS crystalline alumina phase. In many cases, most of the crystalline zirconia may retained as a spongy mass upon the refractory surface, and the expanded volume of alkalialuminosilicate liquid phase simply drain out of it. This expanded volume of liquid phase on the refractory surface then runs down to the glass bath under the force of gravity.

The mechanism written above can be applied in the case of the creation of the cords N° 1 and 2 (Table 3) AZS glassy phase of the refractory located in the melting part above the metal line (Table 6) has practically the same composition like these cords. Cord 1 is enriched by Na_2O, cord 2 by K_2O. We can suppose that an origin of these cords is AZS refractory from the part above the glass melt.

In general, some basic mechanisms of corrosion of refractories by molten glass melt have been recognized [6,13,14]: (1) Diffusion of different species (e.g. Al, Zr, Si) from refractory into glass melts and the species from glass melts (e.g. alkali ions) into refractory; (2) compensation of charge of alkali ions by diffusion of electrons in the material (formation of O_2 bubbles and reduction of multivalent ions present in the glass); (3) erosive action due to the flow of molten glass; and (4) evaporation of volatile components of glass melt, their condensation and diffusion into the material of upper part of the melting furnace.

The bubble forming mechanism appears to be the key for the glass defects originating from fused cast AZS [9]. Other potential source of defects is the formation of a saturation boundary layer when it is combined with an erosive action of the flow of molten glass.

The composition of the cords N° 3-8 (Table 3) is very close to the composition of the AZS refractory glass phase from melting part located below the metal line (Table 7). It implies that bubbles, which have been formed in the AZS glass phase due to compensation of charge of alkali ions by diffusion of electrons in the material, squeezed a part of the glass phase into the bulk of the glass. As there is no boundary layer (Figure 10) the composition of the formed cords is very close to the composition of the AZS glass phase.

All other defects of which compositions are in the Table 3 have probably their origin in the saturation boundary layer present in the glass/refractory interface of working or melting part of furnace. Bubbles, created in the interface, squeezed a part of interface layer into the bulk of the glass or the part of boundary layer has been involved into the glass melt due to the erosive action of the flow of molten glass. This conclusion can be done from the comparison of the boundary layers composition (Tables 8 and 9, Figures 11 and 12) with the compositions of cords N° 9 – 47.

CONCLUSIONS

The study of the relationship between the composition of defects presented at tableware glass products and a glass phase of the AZS refractory material showed that:

- Microstructure of the AZS refractory material located above the metal line differ from the AZS located below the metal line. AZS refractory above the metal line consists only from the ZrO_2 and glass phase, while there are four phases in AZS below the metal line;
- Concentration profiles of individual oxides in the AZS glassy phase above the metal line is practically constant, they increase or decrease in the part below the metal line;
- AZS refractory of melting part and working part located below the metal line differ by the presence of the boundary layer, there is no interface layer in the AZS taken from the melting part;
- AZS glass phase located above the glass melt is characterized by the content of Al_2O_3 about 30 wt. %, while the content of Al_2O_3 is about 22-25 % in the part below the metal line;
- There is a group of cords of which compositions are close to the bulk glass, they have no origin in the AZS material;
- There are the couples of cords with the comparable chemical composition which differ by presence of ZrO_2. A cord coming from the AZS refractory glassy phase may not contain $ZrO_{2;}$
- There is supposed that
  o cords with the content of Al_2O_3 about 30 wt. % come from the part above glass level;
  o cords with the content of Al_2O_3 about 22-25 wt. % originate from the AZS material from melting part located below the metal line;
  o the boundary layer of the melting or working part is the source of the other cords.

REFERENCES

[1] Asokan T.: J. Mater. Sci. Let. 14 (1995), 1323
[2] Fleischmann B.: Glass. Sci. Technol. 78 (2005), 295
[3] Nelson M., Duvierre G., Boussant-Roux, Y.: GlassResearcher 7 (1998), 8
[4] Cabodi I., Gaubil M., Morand C., Escaravage B.: Glass. Tech.: Eur. J. Glass. Sci. Technol. A, October 2008, 49 (5), 221.
[5] Selkregg, K.R., Gupta A.: Ceram. Eng. Sci. Proc. 23 (2002), 59.

[6]Beerkens R.,Van Dijk F.,Dunkl M.:Glasstech. Ber. Glass. Sci. Technol. 77C (2004)35.

[7] Hermans J., Beerkens R.: Ceramics / Silikaty 43 (1999), 147

[8] Dunkl M.: Adv. Mater. Res. 39-40 (2008) 601

[9] Dijk, Van F.A.G.: Glass Defects Originating from Glass Melt/Fused Cast AZS Refractory Interaction. PhD thesis, Eindhoven University of Technology (1994)

[10] Arman B., Aydin E.: Sklar a keramik 50 (2000) 52.

[11] MOTIM Fused Cast Refractories Ltd., Published: February 2012.

[12] Dietzel A.: Glastechn. Ber. 40 (1967), 378

[13] Dunkl M., Bruckner R.: Glasstech. Ber. 53 (1980) 321.

[14] Goris R., Bruckner R., Dunkl M.: Glasstech. Ber. 51 (1978) 294.

ACKNOWLEDGEMENTS

This publication was created in the frame of the project "Industrial research for reaching more effective melting and forming technology of tableware glass" ITMS code 26220220072 and supported by grant VEGA 2/0165/12.

HOW FUSED CAST AZS TAKE CARE OF SECURE AND HIGH QUALITY GLASS MANUFACTURING

Michel Gaubil, Saint-Gobain CREE, Cavaillon, France
Isabelle Cabodi, Saint-Gobain CREE, Cavaillon, France
Jessy Gillot, Saint Gobain Recherches, Aubervilliers, France
Jie Lu, Saint-Gobain Research Shanghai, SGRS, china

INTRODUCTION

Fused cast refractory materials are consider from a long time as the reference for glass contact in soda-lime glass furnaces. Since many years SEFPRO provide refractory solution for this application. Our long experience and knowledge allow us to understand the particular role of AZS in soda-lime glass and to establish the key properties for fused cats AZS fused to secure lifetime and guaranty high glass quality level. Particularly, we will discover how internal structure of AZS soldier block, chemistry - redox and finally microstructure play very important role in glass contact properties

THE KEY ROLE OF AZS REFRACTORY INTERFACE LAYER WITH SODALIME GLASS

Derived from many AZS glass interface study as illustrated in fig 1, we clearly determine equilibrium chemistry at the interface between refractory material and soda-lime glass as illustrated in the curve (fig2) for different temperature

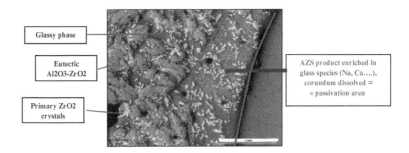

Figure 1: Glass – Refractory interface analysis (1450°C Soda-lime glass)

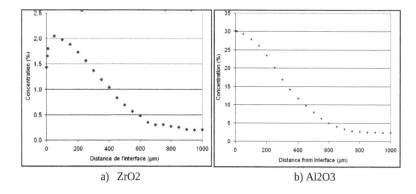

a) ZrO2

b) Al2O3

Figure 2: Glass – Refractory interface analysis (1450°C Soda-lime glass)

Based on these data we succeed melting different synthetic glass representative for the interface composition. We measured at the central Research Laboratory of Saint Gobain Aubervilliers SGR glass properties as glass density and glass viscosity that control glass flow removal at the interface and so corrosion resistance and glass defect ability. The limiting factor for interface corrosion corresponds to diffusion inside the glass. The dissolution will depend on temperature. The measurement results on synthetic interface glasses are illustrated in figures 3 and 4.

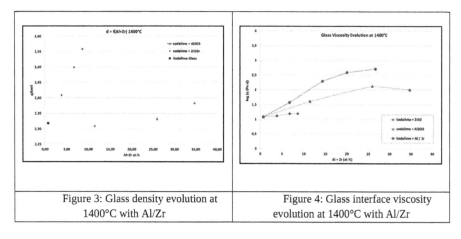

Figure 3: Glass density evolution at 1400°C with Al/Zr	Figure 4: Glass interface viscosity evolution at 1400°C with Al/Zr

The glass density evolution indicate that Al_2O_3 dissolution in soda-lime glass have very limited impact up to 25% on the glass density interface. This interface will by the way only slightly affected by the dissolution of Al_2O_3 from AZS if we consider density driven forces. It will stay stable from this point. Zirconia content has a much greater influence on glass density. Hopefully, the low concentration of ZrO_2 because of limited solubility of ZrO_2 in high alumina glass contributes to limit this effect. Nevertheless these results can explain some corrosion issue with High zirconia fused cast refractory material in soda-lime glass, where horizontal glass refractory interface have high ZrO_2 content with low Al_2O_3 enrichment.

On figure 4, we can observe that Al_2O_3 have a great impact on interface viscosity that will lowered the interface removal. Moreover, we observe a cumulative effect of Al_2O_3 and ZrO_2 enrichment in soda-lime glass. Glass structure arrangement could explain it. We suspect some clustering effect with Al_2O_3 and ZrO_2 in the interface glass layer with four coordinated Al atoms surrounded by Height coordinated zirconium instead of 6 coordinated that limit non bridging oxygen in the glass. The high viscosity results explain why AZS have a good compatibility with soda-lime glass.

Finally, as indicated in figure 5, we measured with the support of Station Sperimentale Del Vetro (SSV) in Murano, surface tension of interface glass composition.

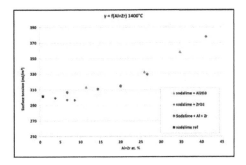

Figure 5: Glass interface surface tension evolution at 1400°C with Al/Zr

We observe in figure 5 that Al_2O_3 enriched soda-lime glass exhibit an increase in surface tension compare to reference. ZrO_2 effect is much limited. This is in accordance with the fact that we can observe, with AZS, noticeable metal line corrosion with marangoni effect that require difference in surface tension between sound glass and interface glass as driven forces to create such corrosion profile. The combined effect of high viscosity increase and low density impact contribute to ensure glass refractory interface stability.

HOW INTERFACE IS AFFECTED BY AZS PROPERTIES

As we better understand the role of glass refractory interface, we will analyses, in the following chapter, how AZS product properties as chemical composition, redox, microstructure size and porosity or residual shrinkage cavity can affect interface and as a consequence corrosion resistance and defect ability.

AZS CHEMISTRY AND REDOX

If we compare similar AZS block from external aspect with similar ZrO2 content as describe in table 1, (See XRF analysis made on block corner) we need to go deeper in the chemical analysis to understand the significant behavior we notice in the following part in between these two AZS blocks.

Table 1: chemical analysis by XRF of AZS A and B

Weight %	ZrO_2	SiO_2	Na_2O	TiO_2	Fe_2O_3	Al_2O_3
Bloc A	33.5	14.6	1.25	0.08	0.08	50.2
Bloc B	32.9	16.5	1.50	0.12	0.11	48.5

In table 1 we see that even these 2 fused cast AZS blocks have similar content of ZrO2, block B exhibits higher silica, sodium oxide and impurities content. That necessary induce higher glassy phase volume, lower viscosity and also higher heterogeneity in the microstructure.

This point is particularly highlighted in the following microstructure analysis in fig6 where we can notice higher glassy phase surface area and more disoriented microstructure for sample B compare to sample A;

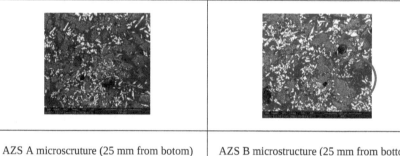

AZS A microscruture (25 mm from botom)	AZS B microstructure (25 mm from bottom)

Figure 6: Microstructure analysis with EDS of AZS A and B

Close to bottom skin, we yet can observe consequences of chemical analysis differences in between A and B sample that are amplified when we perform more detail microstructure analysis from skin to internal part (figure 7)

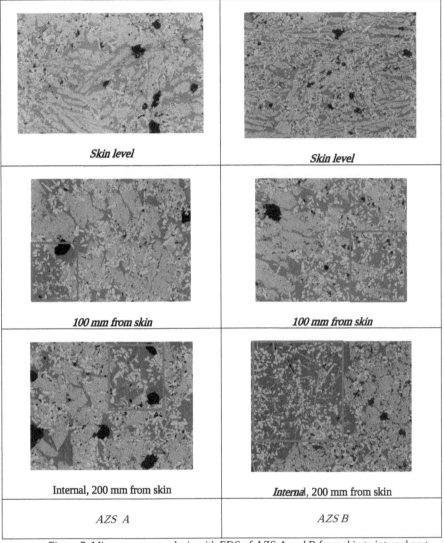

Figure 7: Microstructure analysis with EDS of AZS A and B from skin to internal part

The size and frequency of glassy phase area or "pocket" increase from sample A to sample B due to chemical differences in between them. The following exudation test will demonstrate that the glassy phase viscosity is also affected by the chemistry of AZS block. In the bars exudation test we thermally cycled refractories rods 2 times in between room temperature and 1500°C with 4 hours dwelling time at high temperature for each cycle (see figure 8).

Figure 8: Exudation test device before the test

We can consider that the volume increase of refractories rods in these conditions are representative of glassy phase exudation level, because viscosity at 1500°C during 4 hours maintain glassy phase on the surface of the samples. Nevertheless we place platinum crucible behind the rods to take in account, in case of glassy phase flow down, this additional volume. The results are given in figure 9.

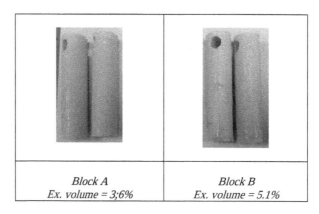

Block A	Block B
Ex. volume = 3;6%	Ex. volume = 5.1%

Figure 9: Glassy phase flowing down the samples

With block B we observe an increase (30%) of ex volume mainly due to higher glassy phase area inside the product and lower viscosity. These microstructure differences are also responsible of different glass refractory interface structure and properties. This is what we will discover behind.

GLASS CONTACT TEST

We performed static corrosion test at 1450°C and 1500°C, during 48 hours in soda-lime glass with same sampling procedure on these two products. The testing device picture is presented in fig 10 where refractory rods are submitted in isothermal conditions to soda-lime glass in platinum crucibles.

Figure 10: Static corrosion test device

As a results, we clearly see in figure 11 the impact in term of glass defect released and interface stability in between AZS A and B, even if similar zirconia content.

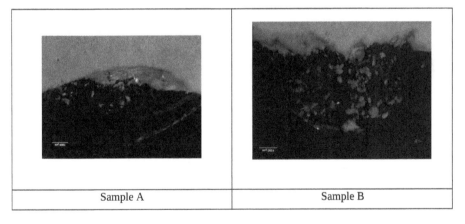

| Sample A | Sample B |

Figure 11: Stoning test results: a) AZS A index 2 , b) AZS B index 4- 5

At 1500°C the stoning index of sample B is more significant than sample A. The microstructure analysis describe in fig 12 below allows understanding, that the higher and heterogeneous interface layers are responsible for the stoning index in the test. The thickness in the case of sample A is close to 200 microns compare to more than 1000 microns in sample B.

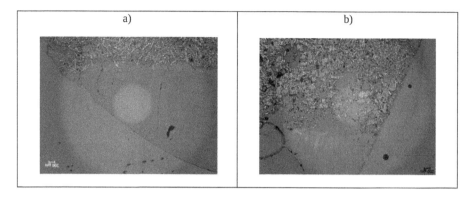

Figure 12: Interface microstructure of static corrosion test: a) AZS A b) AZS B

Combine with lower redox level in block B; we also observe difference behavior in terms of blistering in the transient blistering test. To evaluate blistering ability, we place a disk of refractory material in between two slice of soda-lime glass and put them 1h at 1100°C as illustrated in figure 13. In this simple test we evaluate transient blistering effect who is sensitive to redox aspect.

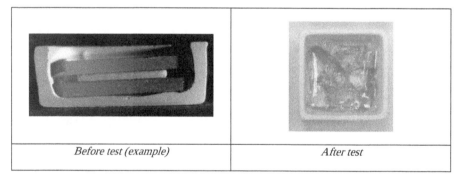

| Before test (example) | After test |

Figure 13: Blistering test in soda-lime glass 1 hour 1100°C
The glass refractory interface observation show very different transient blistering level.

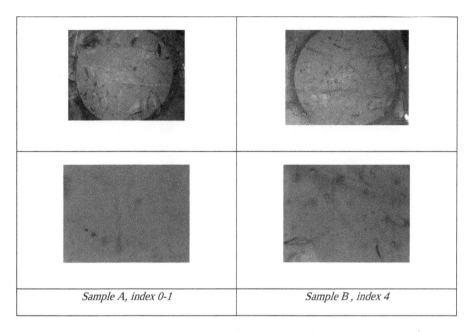

| Sample A, index 0-1 | Sample B , index 4 |

Figure 14: Blistering test results

At interface scale, redox reaction between glassy phase of the refractory and soda-lime glass induce some blisters formation. In bloc B, the higher content of glassy phase combine with higher $Fé_2O_3$ and TiO_2 enhance such reaction. This blistering could have some impact on glass defect but can also initiate some upward-drilling corrosion aspect in fused cast AZS (join area for example).

AZS MICROSTRUCTURE SIZE

At interface level, microstructure characteristics as crystal size play also an important role to control refractory behavior. In order to analyze this point we perform dynamic corrosion test on AZS soldier block ($1500 \times 400 \times 250$ mm^3) as illustrated in figure 15. We draw the future corrosion profile of soldier block in glass furnace by the red curve.

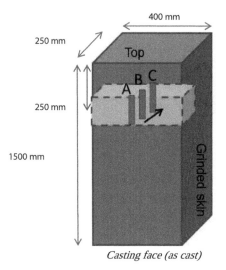

Figure 15: Sampling procedure for dynamic corrosion test

After 48 hours at 1500°C at 6 rpm in soda-lime glass we measure the corroded volume with 3D scan volume analysis before and after test as illustrated in figure 16.

Figure 16: 3D scan analysis of sample after test

The results for sample A, B, and C are given in table 2 in terms of total corroded volume and relative corrosion index. It demonstrate that the skin microstructure as a key effect on corrosion resistance even if the zirconia content at skin is always lower compare to the central sample C. The chemical analysis (XRF data) of the samples is in accordance with chemistry segregation in fused cast AZS soldier block

Table 2: Dynamic test corrosion results and sample chemistry

Sample A	Sample B	Sample C
Corroded volume :6.98 cm3 Index 100	Corroded volume :8.61cm3 Index 81	Corroded volume : 8.82 Index 79
ZrO2 = 36.4, SiO2 =13.70 weight %	ZrO2 = 37.3 SiO2 =12.1 weight %	ZrO2 = 45.1 SiO2 =10.7 weight %

The corrosion is at least 20% higher in the central area despite the zirconia level difference. To understand this phenomenon, we can again observe sample at microscopic scale as in figure 17. We observe microstructure at metal line level on the corroded sample.

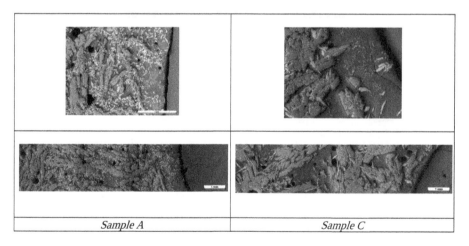

| Sample A | Sample C |

Figure 17: Dynamic test corrosion microstructure analysis

From these glasses – refractory interface analysis, it clearly appears that the size of eutectic alumina-zirconia crystals and the associated size of glassy phase area are significantly different , with more coarse structure in sample C compare to sample B giving to the skin area of the soldier block a significant part of corrosion resistance. For this reason, it is prohibited to make over machining to suppress skin defect aspect of the block to keep the best part of the block. Molding management plays a key role to limit skin layer removal.

MACRO POROITY OR RESIDUAL SHRINKAGE CAVITY

Behind microstructure size, we can consider also that microstructure damage cause by the presence of micro shrinkage cavities at the interface will also have a great impact on corrosion resistance. As illustrated in figure18, we performed dynamic corrosion test at 1450°C during 72 hours 6 rpm, on different sample taken from different area in a regular cast AZS fused cast blocks. The results are shown in figure 19.

Figure 18: Sampling area in fused cast soldier block

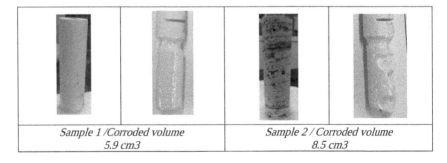

| Sample 1 /Corroded volume 5.9 cm3 | | Sample 2 / Corroded volume 8.5 cm3 | |

Figure 19: Dynamic test corrosion results

We can observe a significant difference in term of corrosion difference between sample 1 and sample 2, with 30% more corroded volume when residual shrinkage porosity are present at the interface. It produces also a particular corrosion profile, signature of the interface heterogeneity.

It point out that the knowledge of the residual cavity position in regular cast fused AZS is a key factor to optimize lifetime of the block. It also point out that re-cutting inside block is not a good practice or need specific rules to prevent any shrinkage cavity at the vicinity of outside surface, close to glass contact.

CONCLUSION

Fused cast refractory material play a key role for soda-lime glass melting thank to the particular glass contact interface. We identify and discuss the interface properties that control refractory behavior in term glass quality and corrosion resistance.

Nevertheless this protective interface layer can be deeply affected by AZS characteristics like glassy phase surface area, viscosity and redox properties. We discover also that crystals size and arrangement, consequences of chemistry and manufacturing process, play a role in corrosion resistance. So, skin structure present a positive role in block lifetime.

This is why, based on this knowledge that SEFPRO built from long time on AZS corrosion process, we establish manufacturing process control such as chemistry, redox control , internal filling (nondestructive control) of the block, microstructure evaluation and machining procedure. This allow us to deliver high quality AZS refractories material that will be guaranty life time and glass quality.

ACKNOWLEDGMENTS

This research was supported by SSV (Statione Sperimentale del Vetro) in Murano for surface tension measurement, SGR (Saint-Gobain Recherches) Aubervilliers for high temperature viscosity and density measurement and also synthetic glass interface melting, finally SGRS (Saint Gobain Research Shanghai) and SG CREE (European Research Center) for refractory testing and chemical analysis.

REFERENCES

1) James Barton and Claude Guillemet, Le verre, science et technologie, EDP Science
2) J Recasens, A. Sevin , M.Gardiol, « comportment de réfractiares au contact de différents veres aux temperature d'élaboration », Verres et Réfractaires , Vol 23, n°1, 1969
3) F Van Dijk PHD Study , "Glass defect originating from glass melt / AZS refractory interaction"

4) G. Tsotridis, E Hondros, « Modelling of the erosion of refractories by marangoni flows", Technische Universitielt Eindhoven , 1998

5) R. Beerkens, "reactions and interaction between tank refractory and glass melt", Norbert Kreidl Memorial Conference , 23-26, june 2004

6) R Bruckner, M Dunkl, J Lexon, "l'influence de différents paramètres sur la corrosion de briques réfractaires par des bains de verres ou de métaux en fusion", SFC , Glast. Ber. 61, n°5, 1988

7) M Boinet, P.Pires Franco, A;Crouzet, L.Massard, M. Gaubil, C Norgot , « localisation of structural heterogeneities in refractory blocks using ground penetration radar », p 141 , Unitecr 2011

ADVANCED FURNACE INSPECTION AND MONITORING BASED ON RADAR SENSORS

Yakup Bayram[1], Alexander C. Ruege[1], Peter Hagan[1], Elmer Sperry[2], Dan Cetnar[2]

[1]PaneraTech, Inc., Chantilly, VA
[2]Libbey Glass, Toledo, OH

ABSTRACT

Furnaces are currently inspected on a regular basis with thermal imaging sensors and other techniques that heavily rely on experience of the plant personnel. However, despite these inspections, the industry still experiences major glass leaks and premature shut-down of furnaces. This results in inefficient asset utilization and major production disruption. Therefore, advanced inspection of furnaces that provides deterministic erosion profile of the refractories and also detects an early stage glass penetration within the insulation layer will result in safer and longer furnace operation through preventive and proactive maintenance.

To address this major industry need, we have been developing radar based Refractory Thickness Sensor and Furnace Tomography Sensors for the last several years. The former measures residual AZS thickness on operational furnace. The latter identifies and 3-D images early stage glass penetration into insulation for preventive and proactive maintenance.

In this paper, we discuss the operational feasibility, case studies and trials for these two sensors, the refractory thickness sensor (RTS) and furnace tomography sensors (FTS). Both sensors fall under PaneraTech's Smart Melter Solution, which combines data from these sensors, the gadgets, with other sensors into a single platform for continuous asset life management and enables preventive and proactive maintenance that are based on deterministic inspection and auditing results.

SMART MELTER SOLUTION

PaneraTech has been developing radar based Refractory Thickness Sensor (RTS) and Furnace Tomography Sensors (FTS) for the last several years. The former measures residual AZS thickness on operational furnace. The latter identifies and 3-D images early stage glass penetration into insulation for preventive and proactive maintenance. These tools have been developed with the user and environment in mind: they survive the harsh glass factory environment and temperatures, are light weight and easy to use to enable quick inspections. These tools are integrated together into PaneraTech's Smart Melter Solution. This solution, represented below in Figure 1, is a combination of these gadgets, with furnace asset management software and a mapping and tracking system, enabling efficient and repeatable inspection for these gadgets.

The combination of the data acquired from inspections using these sensors and brings actionable information to the furnace operators' fingertips. Such data from the sensors can be represented in reports and real-time furnace views. For example, notional drawings of a furnace that has had several inspections completed with the RTS tool is shown in Figure 2. With such visualization, an operator can know very quickly the health of the AZS lining, high risk areas, and the historical progression erosion of the AZS that can be strongly dependent on furnace operations.

117

Figure 1. Smart Melter Solution components

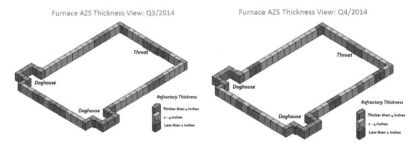

Figure 2. Example refractory thickness risk views of a furnace as would be integrated into the Smart Melter Solution

FURNACE TOMOGRAPHIC SENSOR FOR EARLY STAGE GLASS PENETRATION DETECTION

The furnace tomographic sensor (FTS) maps and identifies early stage glass penetration into the insulation layers backing the refractory lining of glass furnaces. This sensor is based on radar imaging of the insulation layers. PaneraTechhas shown the operational feasibility of several different implementations of this sensor: in-situ scanning, portable scanning, and hand-held quick-scans. All implementations of the sensor is desi gned to fit between structural, cooling, and other elements around the furnace. Tomographic images   of the intern al structure of th e wall are constructed, glass penetrations are automatically identified and mapped. The sensors are designed to work with any sidewall configurations in the glass manufacturing industry.

FTS Feasibility Testing

To demonstrate the fundamental feasibility of the furnace tom ography sensor, we tested the system on a specially constructed developmental furnace at Libbey, Inc. in Toledo, OH. The furnace had four sidewalls consisting of AZS. On one sidewall, a bonded AZS block, super-duty firebrick block and fiber board were placed against the glass-contact fused-cast AZS. The furnace contained molten glass at 2500 °F. The bonded AZS and super-duty block had pre-cut channels to allow glass to f low from a hole drilled in the f used-cast AZS to the outer lay ers. These brick s where surrounded by the same type of brick, but w ithout channels. The test furnace at Libbey is shown in Figure 4, with the scanning FTS system us ed for the feasibility tests. A drawing of the channels cut into th e layers are also shown. A vertical and h orizontal channel guide the molten glass into the horizontal groove cut into the super-duty brick. Mortar was used to seal these blocks to make sure the glass was contain ed in this small area. The installation of the bricks against the fused-cast AZS are also shown in Figure 3.

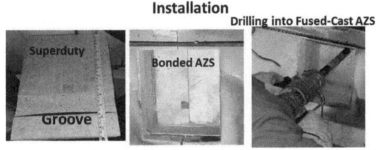

Figure 3. Photographs of the cut-outs into the super duty and bonded AZS bricks and drilling the fused-cast AZS.

We performed the radar tomographic mapping over the area with these cut-outs in addition to the surrounding area which does not contain glass penetrations. This area is shown in Figure 5. A thermal image of the outside of the wall is also shown. There is no indication of glass penetration in this image: the outer surface of the panel is relatively u niform in temperature. After several hours, the sidewall was imaged using the FTS. During this time, the internal furnace temperature was held at 2500 °F.

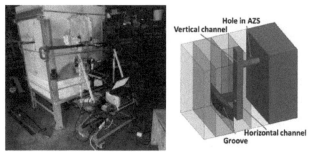

Figure 4. Libbey Test Furnace and diagram of the channels created inside the insulation wall to create glass penetration.

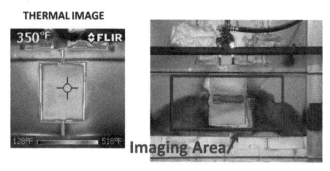

Figure 5. Thermal image of the area with glass penetration and the furnace tomographic system imaging area on the test furnace.

Tomographic images were then created of th e internal wall structure and our autom atic glass detection process was used to identify gl ass penetration at differe nt levels inside the insulation. This glass identification (red) is shown at the fused-cast/bonded AZS interface in Figure 6. As we see, glass is o nly identified in the 10" wide panel and is show n to cover this interface. The glass filling the groove in the super-duty block is also detected.

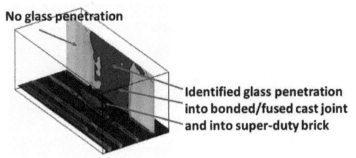

Figure 6. Glass identification at the fused cast/bonded AZS interface joint and into the groove in the super-duty firebrick.

The test furnace was shut-down and after cooling, the sidewall was opened to confirm glass penetration into the channels and in terfaces. Figure 7 are photographs showing that the glass penetrated to the super-duty groove. It was fou nd that glass had filled the interface between th e bonded AZS and fused-cast AZS, as was detected by the FTS. These results prove the feasibility of the sensor technology for glass penetration detection.

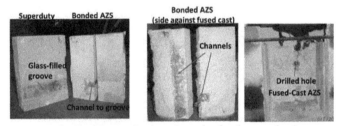

Figure 7. Glass penetration into channels after furnace shutdown.

A second o perational development furnace was constructed in place of the o riginal at Libbey to further assist in the FTS probe development and is shown in Figure8. Glass penetrations at different levels in th e insulation were pre-b uilt into the sidewalls. On one sidewall, glas s penetration into the super-duty level was built in to the walls. The tomographic mapping at this area with molten glass and in intern al furnace temperature of 2600 °F is shown in Figure 8, next to a photograph of the b lock with the cut-out be fore construction and drawing of th e inner wall construction. It is very clear that the FTS can readily capture this structure.

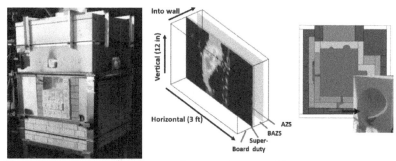

Figure 8. Second Libbey development furnace and tomographic imaging of the cut-out in the second layer (super-duty firebrick) with molten glass inside.

During the tomographic mapping at this sidewa ll, we also discovered an area of glass penetration at the insulation board-super duty interface. No leaks were p re-built into the furnace at this level. A thermal image of this sidewall is shown in Figure 9(a). The temperature at the surface is very uniform. The glass penetration map generated by the FTS is shown in Figure 9(b).

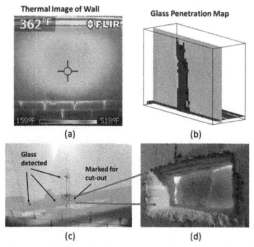

Figure 9. (a) Thermal image of sidewall with suspected glass penetration. (b) Glass penetration map obtained FTS. (c) Glass penetration map marked on sidewall. (d) After the insulation board was removed, revealing the detected glass penetration.

After the glass was mapped using the FTS, the areas of glass penetration was marked on the surface of the insulation board, as shown in in Figure 9(c). A small and specific area was marked for removal of the insulation board which was subsequently and carefully removed. Glass, as predicted by the FTS, was found exactly at the marked location and is shown below in in Figure

9(d). It is very important to note that this area was monitored over time to determine if it was an active glass leak into the insulation board. After monitoring the areas, this glass did not change in area or shape, showing that this was not an active leak.

With these tests at the Libbey development furnaces, we have clearly shown the feasibility of glass penetration detection at deep and s hallow levels inside the insulation package. Additionally, the latest glass detection at the insulation board shown in Figure 9 was truly a blind test. This glass penetration was unexpected and yet the FTS system easily detected this glass.

FTS Blind Test at Container Glass Furnace

Following these tests, we demonstrated the FTS sensor at an operati onal container-glass furnace that was ten years into the furnace campaign. In this blind test, we utilized two versions of the FTS: a hand-held quick-sense FTS used for fast mapping of the whole furnace and a portable scanning FTS used for precision area-mapping of glass penetrations.

We first surveyed the furnace with our hand-h eld FTS, where we walked around the furnace and probed every accessible sidewall area. This entire process took about 2 hours. This tool fits around cooling structur es and buckstays and probes sm all areas. During this initial FTS survey, we identified a location that was of high risk for glass penetration. W e next used our portable scanning FTS to perform imaging of the ar ea. This probe is m eant to be set-up ver y quickly and performs scans between the horizon tal binding steel members on the wall. These scans can range from very small to over 4' in length and completely cover the area between the binding steel members. The portable FTS is designed to be very flexible. It m ounts to existing steelwork on the furnace. It also collapses to a very small package so that long, narrow scans are large, wide and tall scans can be perfor med. The tomographic mapping and glass detection using the portable sensor showed an area of several small glass penetrations at the front, insulation board layer. A photograph of the suspected risk areais shown in Figure 10(a). The portable FTS mounted near the area is shown in Figure 10(b).

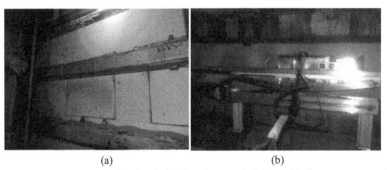

(a)   (b)

Figure 10 (a) Suspect area before insulation board removal. (b) Portable furnace tomographic sensor on sidewall.

We reported the findings to the furnace operato rs, allowing them to access the situation. Temperature measurements on the surface were taken and did no t show abnormal reading s. Subsequently, the insulation board in this area wa s removed. This is shown in Figure 11. Glass was found in this level, as predicted by the FTS. The glass detection along with a ph otograph of the opened area is shown in Figure 12. A large portion of the glass was found along a joint between internal bricks and was also predicted by the FTS. Again, this further proves the FTS as a glass penetration detection system.

In summary, we quickly pin-pointed a high ri sk area after surveying a glass-container furnace using a hand-held version o f the FTS, then performed detailed mapping and finally confirmed the glass detection by opening the ar ea and finding glass as i ndicated in the g lass penetration detection.

Figure 11. Removal of insulation board at glass penetration area

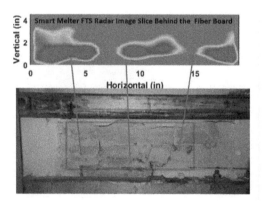

Figure 12. Glass penetration radar mapping of suspect area and photographs of the exposed glass penetration after insulation board was removed.

In-Situ Scanning FTS

The FTS sensor can also be used in an in-situ implementation. This method enables long-term lifetime scanning of the sidewalls. It utilizes the existing binding steel for attachment of the guiding mechanisms along the entire length of the furnace. Opera tionally, the in-situ system consists of a m ounted rail system and a portable sensor box. The m ounted guide-rail system is very low profile and it does not block the view of the furnace si dewalls. The portable sensor box contains the electronics and hardware to obtain the measurements needed for the glass penetration mapping. This portable sensor box is not perm anently mounted on the sidewall and only one is required for a full furnace system with multiple rails. It is simply moved from rail to rail after each scan is done.

We constructed the rail system from low-cost rails where the sensor box was driven by a chain system. The system was mounted on a Libbey furnace, shown in Figure 13. The area marked in red shows the scanning area. There is an abunda nce of steelwork, buckstays, cooling pipes, hoses and other impediments around the furnace. However, the in-situ FTS utilizes the fact that there is a clear area that extends several inches away from the furnace sidewall and spans the entire furnace length to perform very long scans and monitoring of the furnace.

The sensor box mounted on the sidewall guide-rails is shown in Figure 14. It was designed to fit between the sidewall and outer impediments and does not physically touch the sidewall itself. The tomographic mapping of a single depth-cut into the sidewall is shown in Figure 15. This shows the interfaces between each layer in red and is not the glass penetration mapping. No glass penetrations were found in the scan area.

Figure 13. Photograph of furnace with marked scan area, also showing the clearance at the sidewall along the length of the furnace.

Figure 14. The sensor box on the furnace wall at different locations.

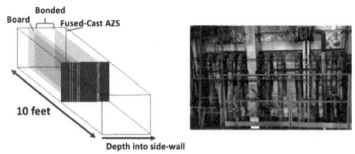

Figure 15. Tomographic results showing the interfaces between the fiberboard and bonded AZS and the bonded AZS and fused-cast AZS near the center of the scan.

In summary, we have shown the operational feasibility of several different implementations of this sensor: in-situ scanning on a real furnace, and portable scanning and hand-held FTS on an operational glass-container producing furnace. These case studies and trials were successful in showing the overall functionality of the FTS for glass penetration detection.

AZS REFRACTORY THICKNESS SENSOR (RTS)

PaneraTech has already developed technol ogy for measuring the AZS refractory lining thickness to the glass, also base d on radar imaging through the refractory lining. This has been implemented in a hand-held solution and an installed, in-situ solution. The hand-held version allows the user to walk around the furnace and pr obe as many spots as possible. C urrently, the hand-held version is ultra-portable. This sm all size and lig ht weight enables the us er to access critical and hard-to-access locations. Additionally, the probe can be used on any type of fused-cast AZS at different Zirconia p ercentages. Ph otographs of its use on operational furnaces are shown below in Figure 16. This version of the tool is weighs less than 2 pounds (0.9 kg) and so is easily mounted on long handles that allow easy access to the exposed metal line on many different furnaces.

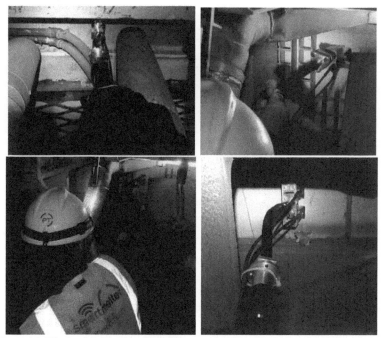

Figure 16. The hand-held refractory thickness sensor in operation at tableware and glass container furnaces.

RTS Sensor Blind Trial at Vidrala Furnace

In this section, we discuss the successful blind trials that we performed at the Gallo-Vidro Vidrala glass factory in Portugal. Measurements at the glass level were performed several days before the furnace d rain and shutd own and were reported to Vidrala befo re the blocks were recovered after the drain. Thus PaneraTech had no knowledge of the thicknesses before reporting the RTS thickness results to Vidrala personnel.

Measurements were performed through fused-cast AZS overcoat blocks (32% AZS) at the metal line. The configuration of the blocks and the m easurement process is shown in Figure 17. We measured through the overcoat blocks at locations where it was suspected that the original block had been completely eroded. Gratings that were holding the overcoat blocks in place were cut prior to measurements to allow for access to the exposed AZS. We performed vertical profile measurements as the figure below s hows in ord er to capture the erosion profile as might be expected at the glass lev el. Eleven areas were measured on many different parts of the furnace. After the shutdown, five blocks were successfu lly recovered. The other blocks could not be recovered due to the construction schedule at the factory. All of the measurements were taken with the furnace in full operation with the glass at nominal, operational level.

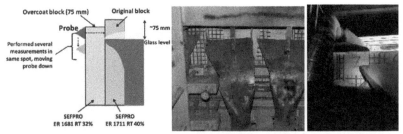

Figure 17. Sidewall configuration and areas at Gallo Vidro that were measured using the refractory thickness sensor

Figure 18 to Figure 20 show the RTS measurement results for spots numbered 5-7, which were located near the furnace thro at. In each figure, we show the act ual locations of the measurements on the cold face of the block, the thickness measured at these locatio ns using the RTS probe, and the final confirmation of the recovered block near the RTS thickness measurement locations. The RTS sensor thickness profile is a vertical scan, starting about 45 mm (1.7 in) from the very top of the overcoat block to 100 mm (4 in) and taking several measurements between.

After the blocks were recovered sh ortly after the furnace drain, it was indeed found that glass was in direct contact with the overcoat blockin these areas: the original block had completely eroded away near the glass level. This was also very evident in the erosion profile observed after the recovery. Note that the thickness measurements obtained from the RTS probe correspond well with the profile seen in the recovered blocks. In these three cases, it is clear we have successfully measured the thickness of the residual AZS using the RTS sensor.

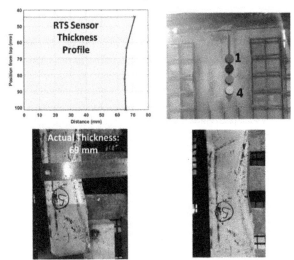

Figure 18. Spot 5 RTS results and measurements after block removal.

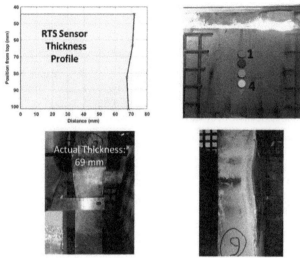

Figure 19. Spot 6 RTS results and measurements after block removal.

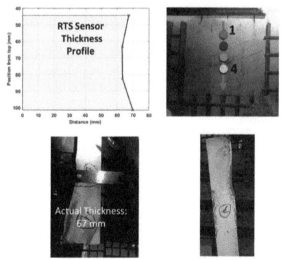

Figure 20. Spot 7 RTS results and measurements after block removal.

A table of the overall results from the trial is shown in Table 1. We record the thinnest area obtained from the RTS measurements and the actual thickness at that area measured after the block recovery. In all five cases, we show an accuracy of 95% or better, when comparing the RTS thickness measurement and actual thickness. These actual thickness measurements were confirmed with Vidrala after the block recovery.

Table 1. Overall Vidrala Spot Measurement Summary

Spot Number	PT's RTS Sensor Reading	Actual Glass Line Thickness	Note
2	73 mm (2.9 in)	73 mm (2.9 in)	Difference: 0mm
3	75 mm (3 in)	71 mm (2.85 in)	Difference 4mm (0.15 in)
5	65mm (2.55in)	69 mm (2.7in)	Difference 4mm (0.15 in)
6	67 mm (2.63 in)	69 mm (2.7 in)	Difference 2mm (0.07 in)
7	63 mm (2.5 in)	67 mm (2.63 in)	Difference 4mm (0.15 in)

Second RTS Thickness Trial

PaneraTech performed an RTS trial at a di fferent glass-container m anufacturer in 2015. As with the Vidrala tr ial, we perf ormed measurements on severa l areas at the g lass level on a furnace to be shortly drained and shutdown. In this case, measurements were through the original glass-contact AZS, which was 34% AZS.

In this case, we were a ble to enter the f urnace after th e drain and reco ver some of the measured blocks. As in the Vidrala trial, Pa   neraTech did not know the thicknesses of the measurement spots during the m easurements. Se veral measurements were again taken ov er a small vertical area in order to capture the erosion wear profile at the glass level. We show one area, seen from inside the furnace in Figure 21. As observed in the photograph, this area shows the deep glass-level erosion cut. These sidewall blocks ha d significant and elaborate erosion profiles. We performed two vertical-scan measurements near this area, the results of which are shown in Figure 22. These two scan measurements were within 6 inches of each other and we found that th e RTS predicted some variation in the profile. The RTS measurement showed a thickness near the glass level of 1.5 inches, where the thickness of the block near the same area was measured to be 1.6 inches. This trial has shown that the RTS sensor can accurately measure the thickness for thinner blocks (less than 2 inch) with complicated geometries, also with glass contact on th e hot-face of the block.

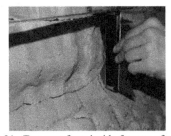

Figure 21. Test area from inside furnace after drain.

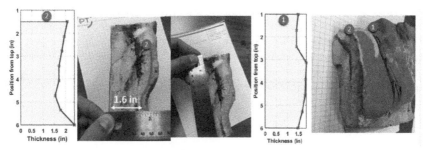

Figure 22. Measured RTS thickness and recovered block thicknesses.

In-Situ Measurements at Libbey Furnace

An in-situ im plementation of the RTS pr obe is ins talled on the furnace itself to continuously and autom atically monitor refractory thickness at particular areas. Two in-situ probes have already been installed at Libbey's Toledo, OH plant on an operational furnace in 2014 and have been in con tinuous operation. The con trol box installed n ear the furnace is shown in Figure 23. The probes are only in contact with the wall during a measurement (which can be less than a second) and automatically retract when not performing measurements. Any number of in-situ probes can be installed in hard-to-reach areas such as throat and in areas wh ere constant monitoring is required. These can be used to monito r critical locations such as the furnace throat and electrode blocks

This in-situ implementation is important for probing critical yet hard-to-reach areas of the glass melting furnace. Currently, we are in progress of installing sensors to continuously monitor the throat refractory thickness at another Libbey furnace, where the throat is very hard to reach.

Figure 23. Control box for in-situ probes installed at Libbey, Toledo, OH.

CONCLUSIONS AND FUTURE WORK

In this paper, we have detailed the RTS and FTS sensors which comprise part of the overall Smart Melter Solution. We have pr oven both sensors through fundam ental feasibility studies as well as operational blind trials at multiple glass-producing furnaces.

Given the compact and flexible nature of these different tools described in this paper, the full sidewall area of a furnace can be m onitored over a scheduled or ad-hoc basis allowing for glass penetration and refractory lining weak s pots to be detected and m onitored over tim e. Integrating the data obtained from these sensors into PaneraTech's Smart Melter Solution, we have enabled deterministic inspection and auditing re sults for continuous asset life m anagement for preventive and proactive maintenance.

Environmental

OPTIMIZING LOW MOMENTUM OXY-FUEL BURNER PERFORMANCE IN GLASS FURNACES TO MINIMIZE FURNACE EMISSIONS AND ALKALI VOLATILIZATION

Uyi Iyoha, Hisashi Kobayashi, Euan Evenson
Praxair, Inc.
39 Old Ridgebury Road, Danbury, CT 06810, USA

Elmer Sperry
Libbey Glass, Inc.
300 Madison Avenue, Toledo, Ohio 43699, USA

ABSTRACT

Glass manufacturers in the U.S. and EU are facing increasing pressure to reduce NOx and particulate emissions from glass furnaces, and to comply with new, lower emissions limits. To meet these ever-tightening limits, glass manufacturers have the choice to install oxy-fuel combustion, which is widely known to produce very low NOx emissions. However, depending on the type of oxy-fuel burners installed in the furnace, the particulate emissions from the furnace and the refractory crown corrosion resulting from increased alkali vapor concentration in the furnace may be excessive. The desire to reduce NOx and particulate emissions, and to avoid rapid crown deterioration have led to the development of low momentum oxy-fuel burners for glass furnaces.

Low momentum burners have been used in oxy-fuel furnaces since the late 1990s. However, commercial experience with these burners shows that depending on the burner design and placement in the furnace, emissions and performance are significantly influenced. To optimize the operation of Praxair's low momentum DOC-WFB burners in oxy-fuel container glass furnaces, a Computational Fluid Dynamics (CFD) model was developed using a commercial glass furnace CFD software package to explore the effects of burner design and placement in the furnace on NOx emissions and alkali volatilization. Our results show that the burner design and elevation of the burner in the furnace, relative to the glass melt, play very important roles in reducing NOx emissions and alkali volatilization. The results show that furnace NOx emissions and alkali volatilization can be reduced by about 30 to 45%, respectively, when the low momentum burner elevation above the glass is optimized. Field measurements of the total particulate emissions (PM$_{10}$) from two oxy-fuel container furnaces fitted with Praxair's low momentum burners measured using EPA Method 5 corroborated this study and showed about a 30% reduction of total particulate emissions for the furnace with higher elevated burners, compared to the furnace with burners placed directly on the tuckstone. In both cases, the NOx emissions remained below 1 lb/ton.

INTRODUCTION

To comply with more stringent NOx limits, oxy-fuel combustion is gaining more and more utilization with over 300 commercial glass melting furnaces being converted to oxy-fuel firing worldwide over the last 20 years. Oxy-fuel furnaces for glass manufacturing offer several benefits including: improved glass quality, reduced furnace footprint, increased specific pull, fuel savings, and reduced NOx emissions. However, depending on the type of oxy-fuel burners installed in the furnace, the particulate emissions and/or refractory crown corrosion resulting from increased alkali concentration may still be excessive. The controlling parameters for silica crown corrosion include the crown surface temperature and gas velocity and alkali concentration at the surface of the crown [1].Thus, optimizing burner design and performance to minimize emissions and alkali volatilization is advantageous and will likely lead to extended oxy-fuel furnace campaigns.

Particulate emissions from glass furnaces are generated by two general mechanisms: physical carryover resulting from high impinging velocity of the flames on the glass surface, and the recondensation of volatile material in the flue gases at lower temperatures [2]. Alkali volatilization in glass furnaces is known to be a complex phenomenon driven by temperature and mass concentration gradients [2]. In addition to increasing particulate emissions, higher alkali volatilization in the furnace has the deleterious effect of increasing the rate of refractory material corrosion due to increased alkali concentrations in the combustion space and in close proximity to the refractory material and furnace crown.

Staged, high momentum flames produce very low NOx emissions. However, because of the high velocities involved, these burners typically result in high rates of alkali volatilization and particulate emissions from glass furnaces. The desire to reduce particulate emissions and rapid crown deterioration led to the development of low momentum, low NOx oxy-fuel burners for glass furnaces, such as burners described in US Patent 6,132,204. Low momentum oxy-fuel burners are burners with flames formed by reacting fuel with an oxidant which has a momentum averaged velocity less than 200 ft/s (preferably less than 100 ft/s) at the exit plane of the gas exit port of the burner. These types of burners have been used in oxy-fuel furnaces since the late 1990s.

These burners are typically installed in the breast walls of the glass furnace, placing the oxy-fuel burner blocks directly on the tuckstone of the furnace. At this elevation, the plane of the flame sheet is on the order of 8 to 18 inches above the glass melt. However, despite the burner being installed horizontally in the glass furnace and the flame starting out parallel to the glass melt, we have observed that when such low momentum burners are operated in such close proximity to the glass, the flames are drawn towards the glass melt (Figure 1) as a result of the Coanda effect, which is the tendency of a fluid jet to be attracted to a nearby surface. This phenomenon affects the performance of the oxy-fuel burner, increasing alkali volatilization, particulate, and NOx emissions.

Figure 1. In-furnace side view of low momentum oxy-fuel burner placed directly on the furnace tuckstone, showing soot-laden upper region and flames pulled downwards towards the glass melt due to the Coanda effect.

One approach to reduce alkali volatilization and carryover is to minimize the velocity of the flame in close proximity to the glass melt. This can be achieved by increasing the elevation of the oxy-fuel burner, relative to the glass melt [3]. However, simply increasing the burner elevation with a conventional burner significantly increases the temperature of the furnace crown, resulting in refractory overheating, rapid deterioration and expensive maintenance costs. To address the increase in crown temperature, one approach that was previously considered and commercialized is the Tall Crown Furnace concept [3] which elevates the furnace crown. This approach has been commercially demonstrated to successfully reduce alkali volatilization. However, this approach also results in increased refractory cost and increased wall heat losses. Thus, lower crown height is preferred to minimize the additional expense and adverse impact on energy efficiency.

Using a dilute oxygen combustion, low momentum wide flame burner (DOC-WFB), the radiation of the flame is preferentially intensified below the flame and directed towards the glass melt (Figure 1), while the sub-stoichiometric upper region of the flame produces a sooty, optically dense cloud which minimizes the direct radiation from the flame to the furnace crown. This phenomenon allows the elevation of the DOC-WFB higher off the glass melt and towards the crown, without appreciably increasing the temperature of the furnace crown.

MODELING APPROACH

To evaluate the effect of burner design and position on NOx emissions and volatilization, a three-dimensional computational fluid dynamic (CFD) model of the oxy-fuel furnace was developed using the commercially available Glass Service GFM package. For this study, two burner types were considered — a high momentum, ultra low NOx DOC-JL burner and a low momentum, medium NOx, wide flame burner, DOC-WFB. The furnace modeled was a flint, container glass furnace operating at a pull rate of 388 stpd, glass melting area of 117.8 m². The burners were modeled at different elevations in the furnace, starting with the burner placed directly on the tuckstone (Elevation 1), and then raised sequentially from Elevation 1 to Elevation 4.

RESULTS

The CFD model was used to evaluate the effect of burner design and placement on the flame characteristics, interaction with the glass melt, furnace temperature, NOx emissions, volatilization, and alkali concentration at the furnace crown. The results of the model study are presented in several 2D slices through the combustion space. For the furnace with high momentum burners (Figure 2a) and low momentum burners (Figure 2b), 10 and 8 burners arranged in staggered configuration in the breast walls were modeled in the furnace, respectively. As a result of the wider flame profile of the low momentum burners, fewer burners are required to achieve similar flame coverage (Figure 2), heat transfer to the glass, and temperature profile. The ability to use fewer of these low momentum wide flame burners, relative to conventional oxy-fuel burners, is advantageous for oxy-fuel furnace operation as it reduces the CAPEX requirement for the oxy-fuel flow control equipment by about $50,000 to $100,000, depending on furnace size.

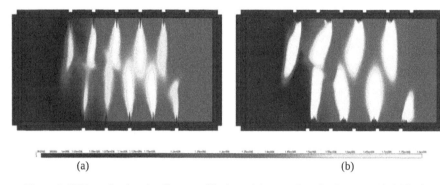

(a)   (b)

Figure 2. CFD results showing flame profile through burner plane for furnace with (a) high momentum DOC-JL burners and furnace with (b) low momentum DOC WFBs. Burners installed on the tuckstone - Elevation 1.

Figure 3 shows the temperature profile of a slice of the combustion space for the low momentum burner. Consistent with the observations in actual commercial glass furnace operations, the CFD result in Figure 3 shows that when the low momentum burner is installed directly on the tuckstone, the flame has a tendency to be attracted towards the glass melt as a result of the Coanda effect. The implication of this is that despite being a low momentum burner, the attraction of the flame towards the glass surface results in the flame impinging on the melt, causing local hot spots and high velocity regions which appreciably increase alkali volatilization. The interaction of the flame and the glass melt also decreases the entrainment and dilution of the jet by products of combustion present in the combustion space, resulting in higher peak flame temperatures and higher NOx emissions. By contrast, however, when the burner is elevated higher in the furnace breast wall (Figure 4), the Coanda effect diminishes and the flame profile is observed to be straighter and higher above the glass surface, reducing the interaction of the flame with the glass.

Figure 3. CFD results showing temperature map of slice through the low momentum burner installed directly on the furnace tuckstone (Elevation 1).

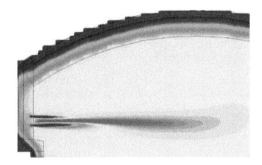

Figure 4. CFD results showing temperature map of slice through the low momentum burner for burner elevated higher off the furnace tuckstone.

As previously mentioned, one of the limitations preventing the higher elevation of conventional oxy-fuel burners in glass furnaces is the potential to adversely increase the crown temperature. Figure 5 shows the effect of elevating the low momentum burner on the average furnace crown temperature. The data plotted in Figure 5 correspond to the case in which the low momentum burner sits directly on the tuckstone (Elevation1) and the case where the burner is elevated higher off the glass surface (Elevation 3). The graph shows that the average crown temperature only increases by about 7°C at around the furnace hotspot, while there is almost no change in temperature over the rest of the furnace crown. The reason that only a minor increase in crown temperature is observed, as opposed to more severe crown temperature rise with burner elevation, is due to the improved radiative properties of the WFB design.

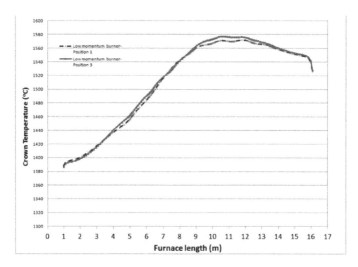

Figure 5. Average crown surface temperature from CFD model for oxy-fuel furnace with low momentum burner placed on the tuckstone (Elevation1) and elevated higher off melt (Elevation3).

A plot of the instantaneous flame speed on the glass surface (Figure 6), calculated by computing the velocity magnitude of the combustion space just above the glass melt in the CFD model, illuminates the benefits of elevating the low momentum burners. As previously discussed, one approach to reduce alkali volatilization and carryover is to minimize the velocity of the flame impinging on the glass melt. Figure 6 shows the average velocity magnitude for the low and high momentum burners at the different elevations of this study. The graph shows that, compared to Elevation 1 at which the burner is located directly on the tuckstone, when the burner is elevated higher off the glass surface to Elevation 3, the average velocity magnitude at the surface of the glass dropped by 61% and 35% for the low momentum and high momentum burners, respectively. By comparison to the high momentum burner placed directly on the tuckstone (Elevation 1), the data shows that the low momentum burner has the potential to reduce the flame impinging velocity on the glass surface by as much as 70%. It is interesting to note that when the burner is further elevated from Elevation 3 to Elevation 4, the average velocity magnitudes for the two burner types are observed to increase, compared to Elevation 3.

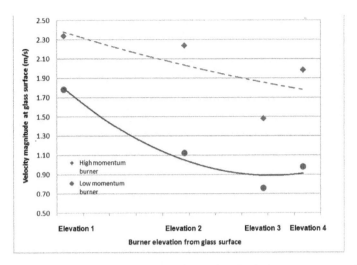

Figure 6. Average velocity magnitude at the glass surface for the low momentum and high momentum burners as a function of burner elevation.

The average temperature of the glass surface (Figure 7) was also estimated from the CFD model by averaging the bottom 7 cm of the combustion space temperature. Interestingly, the analysis showed that when placed directly on the tuckstone (Elevation 1), the low momentum burner resulted in higher average glass surface temperature compared to the high momentum burner. This is consistent with Coanda effect drawing the flame towards the glass melt and impinging on the melt surface. Impinging flames create localized regions of high gas impact velocities and elevated glass surface temperatures. Also, as previously discussed, the impinging flame on the glass melt decreases the entrainment and dilution of the gas jet, resulting in higher, local peak flame temperatures. The graph shows that as the low momentum burner is elevated, the average glass surface temperature drops.

By comparison, the high momentum burner is only observed to show minimal decrease in the average glass surface temperature as the burner is elevated. The CFD results show that the high momentum burner does not experience the same influence of the Coanda effect pulling the flame towards the glass surface, when the burner was placed directly on the tuckstone.

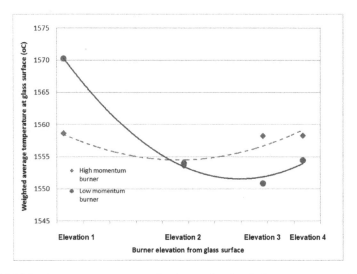

Figure 7. Weighted average temperature at the glass surface for the low momentum and high momentum burners as a function of burner elevation.

Figure 8 shows the NOx emissions calculated by the CFD model for the low momentum and high momentum burners at the different elevations in this study. The NOx results for the burners at Elevation 1 are in relatively good agreement with data generated for these burners in Praxair's combustion laboratory, which shows that the high momentum burners produce about 60% less NOx than the low momentum burners at the same furnace operating temperature and nitrogen concentrations. However, when the burners are elevated in the CFD model, the NOx emissions data is observed to depart from the laboratory trend.

The CFD results show that when the low momentum burner is elevated from the tuckstone, there is a significant reduction in NOx emissions as a function of burner elevation. At the optimal elevation corresponding to Elevation 3, the NOx emission is observed to decrease by 31% (Figure 8). This reduction is likely the result of the lifting of the flame from the glass surface, resulting in higher entrainment of furnace gases by the jet, diluting the oxygen and natural gas stream and reducing the peak flame temperature. This is evidenced by the reduction of the average temperature just above the glass shown in Figure 7. By comparison, however, the NOx emissions for the high velocity burners (Figure 8) is observed to be only slightly decreased when the burner is elevated to Elevation 2, and actually increase by as much as 25% when the burner is further elevated. The observed increase in NOx emissions is in good agreement with the increase in the weighted average temperature of the combustion space captured by the CFD model as the high momentum burner is elevated above Elevation 2 (Figure 7).

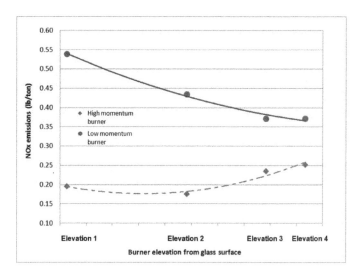

Figure 8. CFD modeling results of NOx emissions for the low momentum and high momentum burners as a function of burner elevation.

Figure 9 shows the alkali (NaOH) emissions from the flue port calculated by the CFD model for the low momentum and high momentum burners, as a function of the different burner elevations considered in this study. Interestingly, despite being a low momentum burner, the results show that the NaOH emissions for the burners are very similar for the base case in which the burners are placed directly on the tuckstone. A plausible explanation for the high alkali emissions for the case in which the burners are placed directly on the tuckstone (Elevation 1), despite the lower velocity magnitude shown in Figure 6, may be the higher glass surface temperature observed at this elevation. As is well known, volatilization rate increases sharply with glass surface temperature due to the exponential increase of the equilibrium vapor pressure [5].

Similar to the effect on NOx emissions, a significant reduction in alkali emissions is observed for the low momentum burner with increasing elevation. At the optimal location corresponding to Elevation 3 of this study, a 45% reduction in NaOH emission from the flue port is observed for the low momentum burner when the burners are elevated from Elevation 1 to Elevation 3. The significant reduction in emissions is in good agreement with the reductions in velocity magnitude and average glass surface temperature shown in Figure 6 and Figure 7, respectively. By comparison, however, at Elevation 3, the alkali emissions in the flue port for the high momentum burner is only observed to decrease by a maximum of about 12%, compared to Elevation 1. The data shows that at the optimal location, the low momentum burner has the potential to decrease alkali emissions by as much as 47% compared to conventional high momentum burners placed directly on the furnace tuckstone.

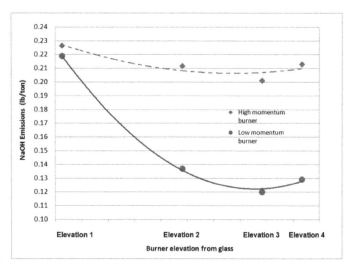

Figure 9. CFD modeling result of NaOH emissions from oxy-fuel glass furnace for low momentum and high momentum burners as a function of burner elevation.

Figure 10 to Figure 13 show 2-D plots of the contour lines and alkali concentration (ppm, mass) map through the furnace centerline for the high and low momentum burners at the different elevations of this study. Figure 10 and Figure 11 correspond to the burners placed directly on the tuckstone (Elevation 1) for the high and low momentum burners, respectively. As shown in the concentration map, the alkali content in the combustion space is very similar for both the high and low momentum burners, indicating that at this burner elevation, despite the low momentum design, the low momentum burner results in about the same level of volatilization as the high momentum burner. This is consistent with the data plotted in Figure **9**, which shows that the alkali emissions at the flue port for both burners were almost equal. By contrast, however, when the burners are elevated, an appreciable reduction in alkali volatilization is observed for the low momentum burner.

Figure 12 and Figure 13 correspond to the alkali concentration maps for the cases where the burners are raised to Elevation 3. For the low momentum burner, the peak NaOH concentration in contact with the crown is 371 ppm by mass, while for the high momentum burner at the same elevation the peak NaOH concentration in contact with the crown was calculated to be 686 ppm, mass. By comparison to the low momentum and high momentum burners at Elevation 1, this observed decrease in alkali concentration at the crown surface attained by elevating the low momentum burners corresponds to about a 46% reduction.

The implication of the ability of the Praxair low momentum burner to significantly lessen alkali volatilization and concentration at the furnace crown is the potential to reduce crown corrosion and furnace repair expenses, and the possibility of appreciably extending the length of the oxy-fuel furnace campaign.

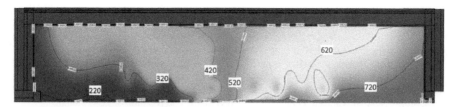

Figure 10. NaOH concentration map (ppm, mass) for oxy-fuel glass furnace with high momentum oxy-fuel burner placed directly on tuckstone (Elevation 1).

Figure 11. NaOH concentration map (ppm, mass) for oxy-fuel glass furnace with low momentum oxy-fuel burner placed directly on tuckstone (Elevation 1).

Figure 12. NaOH concentration map (ppm, mass) for oxy-fuel glass furnace with high momentum oxy-fuel burner raised to Elevation 3.

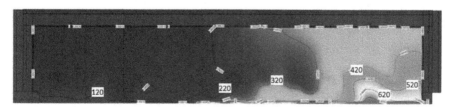

Figure 13. NaOH concentration map (ppm, mass) for oxy-fuel glass furnace with low momentum oxy-fuel burner raised to Elevation 3.

The findings from this CFD modeling work were recently adopted and used in the burner design and placement for a 200 stpd flint oxy-fuel container glass furnace. In that furnace, 6 Praxair DOC-WFBs were installed and elevated higher above the tuckstone. A comparison of the total particulate emissions (PM_{10}) from the new furnace measured using EPA Method 5, with the total particulate emissions from a separate furnace with 6 similar DOC-WFBs but installed directly on the furnace tuckstone showed that about 30% reduction of total particulate emissions was achieved (data was corrected for electric boosting) using the low momentum DOC-WFB technology. NOx emissions in both cases were less than 1.0 lb NOx/ton of glass pulled.

SUMMARY AND CONCLUSION

To comply with more stringent NOx limits, oxy-fuel combustion is gaining widespread utilization with over 300 commercial glass melting furnaces being converted to oxy-fuel firing worldwide. A CFD model was developed using a commercial glass furnace CFD software package to explore the effects of burner design and placement in the furnace on flame characteristics, NOx emissions and alkali-volatilization and concentration at the crown surface. Two types of burners were considered for this study – a low momentum DOC wide flame burner and a high momentum DOC J-L burner.

The CFD results for the low momentum burners placed directly on the furnace tuckstone were consistent with actual field observations in commercial furnaces and showed that when the low momentum burners are installed directly on the tuckstone, the flames have a tendency to be attracted towards the glass melt as a result of the Coanda effect, resulting in the flame impinging on the melt surface and resulting in local hot spots and high velocity regions which appreciably increase alkali volatilization and NOx emissions. By contrast, however, when the low momentum burners are elevated higher in the breast wall, the Coanda effect was diminished and the flame profile was observed to be straighter and higher, reducing the interaction of the flame with the glass surface.

The results show that by using the low momentum burner and positioning the burner at an optimal elevation in the furnace, the NOx emissions, alkali volatilization, and alkali concentration in close proximity to the furnace crown can be decreased by 31%, 45%, and 46%, respectively, when compared to the base case low momentum burner placed directly on the furnace tuckstone.

These findings were recently implemented in the design of a 200 stpd flint oxy-fuel furnace. Comparison of the total particulate emissions (PM_{10}) from the new furnace measured using EPA Method 5, with emissions from a furnace with 6 similar DOC-WFBs but installed directly on the tuckstone showed that about 30% reduction of total particulate emissions was achieved using Praxair's low momentum burner technology. NOx emissions achieved in both cases were below 1.0 lb NOx/ton of glass pulled.

REFERENCES

1. W. Snyder, K.T. Wu., "Reduction of alkali volatilization and refractory corrosion for oxy-fuel fired furnaces," XV A.T.I.V. Conference, Parma, Italy, 1999.
2. Benjamin Jurcik et al., "Particulate Emissions in Oxy-Fuel Fired Glass Furnaces," Ceram. Eng. Sci. Proc., 17 [2] 36-46 (1996).

3. H. Kobayashi, K.T. Wu, G.B. Tuson, F. Dumoulin, H. Kiewall, "TCF Technology for Oxy-Fuel Glassmelting," Refractories, 2005.
4. W. Snyder, A. Francis, "Wide flame burner," U.S. Patent 6,132,204 A.
5. H. Kobayashi, "Advances in Oxy-Fuel Fired Glass Melting Technology," XX International Congress on Glass (ICG), Kyoto, Japan, 2004

HEAT OXY-COMBUSTION: AN INNOVATIVE ENERGY SAVING SOLUTION FOR GLASS INDUSTRY

Hwanho Kim, Taekyu Kang, Kenneth Kaiser, Scott Liedel
Air Liquide R&D, DRTC, Newark, Delaware, USA
Luc Jarry
Air Liquide Metal and Glass Market, Shanghai, China
Xavier Paubel
Air Liquide ALTEC, France
Youssef Jumani
Air Liquide R&D, Paris-Saclay, France
Levent Kaya
SISECAM R&D, Istanbul, Turkey

ABSTRACT

Oxy-combustion today is a mature technology and well known for higher efficiency and lower NO_x compared to air combustion even without recovering waste energy from the flue gas. Rising energy cost and more stringent environmental regulation make oxy-combustion technology more attractive. In order to take it one step further, HeatOx uses waste energy from the flue gas to preheat oxygen and natural gas to 650°C and 450°C, respectively. The technology employs air as an intermediate heat transfer fluid instead of direct heat exchange between flue gas and O_2/NG to guarantee a safe and maintenance free operation. HeatOx is a proven technology with two float glass furnace references, where 10% fuel reduction is demonstrated compared to the traditional oxy-combustion and by up to 25% compared to the typical regenerative air-fired furnaces.

A collaborating team is implementing a new HeatOx system at an industrial scale. The main objective is to reduce CO_2 and NO_x by 23% and 90% compared to traditional air combustion for tableware glass production. The HeatOx technology described here is specifically adapted to container, fiber and technical glass furnaces. Special attention is given to a staged and compact burner and multi-channel oxygen and natural gas preheaters. The burner is operable without intervention with both hot and cold reactants maintaining constant flame coverage. The multi-channel preheaters feed oxygen and natural gas to multiple burners with independent control of flow rate and temperature for each burner.

INTRODUCTION

Reducing energy consumption and minimizing pollutants emissions have been two major challenges in the glass industry. Fuel consumption has decreased from 12.5 GJ/Ton to 4.5 GJ/ton in the container glass sector since the 1960's [1]. During the same time, NO emissions in the atmosphere have been reduced down to 1500 mg/Nm^3 in air-fired float furnaces [2]. Oxygen combustion technology has contributed to reach these targets. For example, more than 50% of reinforcement fiber glass furnaces have been converted to oxy-combustion in Europe. Heat recovery is also an important technology to improve energy efficiency and reduce emissions. Energy can be recovered from the flue gas or from any fluid (blowing air, hot water) and be used usefully in the process instead of wasting it. Cullet preheaters or Batch & Cullet preheaters are a great example of heat recovery. Typically, from 10% to 15% savings are reported depending on the cullet ratio, preheat temperature and moisture content. Other technical solutions, such as thermo chemical recuperation (TCR), Organic Rankine Cycle, waste heat boiler have also been studied in the glass industry for several years [3].

We have developed oxygen combustion technology for the glass industry for decades. To further improve the energy efficiency, we have integrated a heat recovery technology into oxy-combustion systems by preheating oxygen and fuel (natural gas). In the float glass sector, we have demonstrated that fuel consumption can be reduced by 25% and NOₓ by 83% when converting an air-fired furnace to an oxygen fired furnace with a heat recovery system [4]. Preheating oxygen and natural gas is one of the promising technologies to increase efficiency in glass melting tanks [5]. When both gases are heated to 500°C, approximately 10% energy efficiency gain can be expected. To achieve fuel savings and emission reductions at the same time, the SUN burner, a staged oxyburner, has been developed. It has multiple burner blocks allowing extreme staging of oxygen and achieving very low NOₓ emissions independent of the oxygen purity, nitrogen content in the natural gas and volume of air ingress. The SUN burner with the heat recovery system has been shown to reduce fuel consumption by 10% and NOₓ emissions down to 100 ppm.

In order to further extend the success of heat recovery integration with oxy-combustion system in the flat glass sector, Air Liquide continued to develop an innovative Heat Oxy-combustion (HeatOx) solution. including newly developed burners and preheaters for small and medium scale furnaces, between 100 tpd to 300 tpd, such as the reinforcement fiber, insulation fiber, container or tableware glass sectors. HeatOx burners are more compact than SUN burners, so they fit within the furnace superstructure requirements. New compact preheaters for natural gas and oxygen have been specifically designed for these furnaces requiring lower capital costs and a smaller footprint. In a partnership with SISECAM, supported by LIFE European program, the technology is being installed at one of the glassmaker's plants. This paper will present the HeatOx burner design features, the multi-channel oxygen and natural gas preheaters operation philosophy as well as the pilot scale demonstration results and industrialization overview of the technology.

HEATOX BURNER

A FC burner features a staged oxygen combustion burner making wider flame compared to traditional pipe-in-pipe burners. By design, the FC burner generates less NOₓ emissions (50 - 100 mg/MJ) than pipe-in-pipe burners (100 to 250 mg/MJ) by controlling oxygen dilution and peak flame temperature. FC has been widely installed in all glass sectors: float glass, container, borosilicate, reinforcement fiber. In order to integrate the heat recovery technology into oxy-combustion burners, FC burner has been chosen as a baseline of HeatOx burner development due to its compactness and wide use in the industry. Figure 1 shows a general concept of burner. The burner uses a patented design where fuel and oxygen mix outside the burner block. Fuel and oxygen are introduced into the furnace through a unique configuration of injectors that produces a highly luminous flame up to three times wider than conventional pipe-in-pipe oxy-fuel burners. The fuel is distributed, through a refractory burner block, among three injectors arranged horizontally, so that the streams of fuel form a wide blanket of fuel in the furnace.

Figure 1. HeatOx burner concept (left) and the flame from the pilot scale test (right)

With the basic concept, HeatOx required significant design development to utilize both hot and cold reactants without intervention. The burner not only operates safely with high temperature natural gas and oxygen, but the burner performance (heat transfer and emissions) stays the same compared to FC burner. Materials for the burner and HeatOx system have been selected carefully based on long term oxidation studies in order to alleviate the safety issues regarding high temperature oxygen and natural gas. The burner can be operated with natural gas at any temperature from ambient to 450°C. Oxygen can be introduced at ambient temperature or any temperature up to 650°C. Several weeks of trials have been performed to validate the following features of the burner: Flame length, flame width, flame efficiency, NO_x emission, temperature profile, pressure drop in natural gas line. Two versions of the burner have been tested: 1MW and 2MW in both cold (oxygen and natural gas at 25°C) and hot (oxygen at 450°C and natural gas at 450°C) configurations. Figure 2 shows the root flame shape for different powers for a 2MW HeatOx burner.

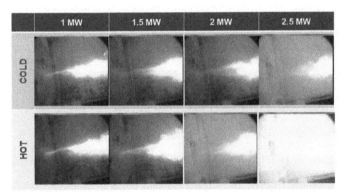

Figure 2. Flame luminosity as function of burner power and reactant temperature (HOT: oxygen and natural gas at 450°C, COLD: oxygen and natural gas at 25°C)

It should be noted that a significant velocity increase is expected when the reactant is heated due to its volumetric changes, which affects the flame stability, shape, heat transfer, and emissions. When the same gas nozzle is used, the average jet velocity of reactants increases about 2.5 times at 450°C. The burner features a patent pending smart device to compensate this natural gas volume changes and have consistent jet velocity and momentum between cold and hot reactants. It is a reversible and passively acting device depending on the reactant temperature. Thanks to this innovative device, reactant temperature did not make any noticeable changes to the visible flame length. In the visible flame (mostly due to soot emissions), flame lengths appeared to vary from 2.5 m at 1MW to 3 m at 2.5 MW when both reactants were preheated. The flame with cold reactants still maintained its shape, roof temperature, and heat transfer profile. Flame width varied from 0.8 m to 1.2 m as power decreases. The flame was very luminous making high concentrations of soot which is very suitable for glass melting as radiation emits in the range (400 – 2500 nm) ideal for penetration in the melt.

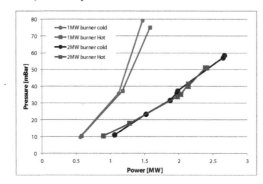

Figure 3. Pressure evolution for the burner as function of burner power

Pressure drops across the burner have been measured at the inlet of the natural gas lance and at the inlet of the oxygen header. Figure 3 shows pressure evolution of the natural gas lance for the two versions of the burner (1MW and 2MW) in terms of various powers at cold (ambient) and hot (450°C) reactant temperature. Natural gas pressure did not exceed 100 mbar regardless of the burner configuration. Thanks to the smart device, the pressure drop of natural gas (Figure 3) was almost identical between cold and hot reactants. Oxygen pressure drop remained below 80 mbar, even when the reactants were preheated. The pressure drop was measured at stable operating conditions. The pressure fluctuation was minimal during the change of burner power and the transition of reactant temperature.

The oxygen staging ratio and nitrogen contents in oxygen and natural gas affects NO_x emissions. The burner can control the oxygen staging ratio by changing an orifice in the burner. NO_x emissions were also evaluated at different staging ratios and reactant temperatures during the pilot scale tests. Figure 4 shows measured NO for different configurations in flue gases at about 3% dry O_2. NO emission increases with the addition of N_2 for both cold and hot reactants up to about three times at maximum N_2 flow (10% of NG by volume). NO emission increases nonlinearly with the secondary orifice diameter (area opening). The change in NO is minimal between 10% and 15% area opening for cold reactants, but it suddenly increases about 150% when the orifice area opening reaches 20%.

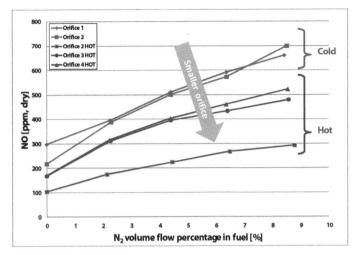

Figure 4. Burner NO measurements in pilot furnace
(HOT: oxygen and natural gas at 450°C, COLD: oxygen and natural gas at 25°C)

The burner is primarily designed to operate at hot reactant conditions, where the oxygen jet has an optimal velocity and momentum. This is optimized to have less NO with hot reactants than the cold reactants case. For the same size orifice, the burner generates less than 50% of NO emission with hot reactants compared to cold reactants. The present results indicate that NO can be reduced as low as 100 ppm upon the selection of staging ratio for natural gas containing very low nitrogen. Even for 5% nitrogen in the fuel, NO level can be as low as about 200 ppm. At the industrial scale, the oxygen staging ratio (orifice selection) will be carefully optimized in order to achieve the efficiency increase and emission reduction simultaneously.

OXYGEN AND NATURAL GAS PREHEATER

In parallel to HeatOx burner development, Air Liquide has worked on the development of a complete system for preheating oxygen and natural gas. Newly developed patent pending preheaters preheat the reactants safely through a multi-channel design, which allows feeding multiple burners and controlling the gas temperature individually. Natural gas and oxygen preheaters equipped with two independent lines for feeding preheated reactants to two burners were demonstrated at pilot scale. Figure 5 shows a picture of the pilot demonstration system. The furnace is coupled with a fumes/air heat exchanger, a primary heat exchanger. The air is preheated between 600°C to 700°C through the primary heat exchanger. This hot air (heat transfer fluid) delivers the recovered energy from flue gas and transfers the energy for preheating oxygen and natural gas. This demonstrated system is a downscaled version of the one to be installed at a SISECAM plant in Europe.

Figure 5. Pilot scale burner system for indirect preheating of
oxygen and natural gas by flue gas recovered heat

INDUSTRIALIZATION OF THE TECHNOLOGY

The burners have been operated in one of the oxygen-fired furnaces, producing 200 tpd
of glass. Figure 6 shows an overview of burner configuration and how the burner equipment is
schematically installed around the furnace.

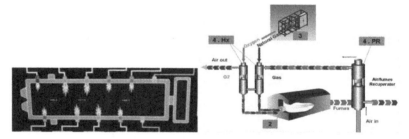

Figure 6. Fire distribution of SISECAM oxy-fired furnace with
FC burners (left) and schematic view of HeatOx solution (right)

Impact of oxygen and natural gas preheating has been proven for the last ten years:
Unfavorable impact on the process (refractory corrosion, batch and glass properties, dust, and
evaporation) has not been observed. Identified benefits are 90% reduction in NO_x emissions and
23% reduction in CO_2 and fuel savings of 23% compared to air fired reference. In September
2015, the HeatOx system was commissioned and started recovering the heat from the flue gas.
The additional benefits of HeatOx system over cold oxygen system will be addressed as
accumulating more furnace operational data.

CONCLUSION

After the first reference in float glass domain that has demonstrated substantial benefits on NO_x emissions and fuel consumption, the burner has been tailored for medium and small size furnaces used in tableware, container and fibers sector. The burner system has been tested and validated for oxygen and natural gas temperature up to 450°C in a pilot scale. Multiple burners can be simultaneously operated with independent power control and fast response with load (reactant flows) variations. The key features of the burner are following:

- Compact and operable with hot and cold reactants
- Constant flame length (~3m) with hot and cold reactants.
- NOx level can be controlled under 200 ppm.
- Flame luminosity enhancement.

Industrial scale burner system has been commissioned in 2015 at one of the plants in Europe.

ACKNOWLEDGMENT

Air Liquide and SISECAM thanks European commission for supporting Eco-Heatox project through its LIFE funding program.

REFERENCES

1. Van der Woude, J. H. A. (2013), "is 50% energy efficiency improvements in glass Product/production chain feasible – Needs of the NL glass industry for the future", Glass trend-ICG Seminar & Workshop on Innovation in Glass Production, Eindhoven, The Netherlands.

2. Scalet, B. M., Garcia Munoz, M., Sissa, A. Q., Roudier, S., & Delgado Sancho, L. (2013), "Best available techniques (BAT) reference Document for the Manufacture of Glass", Industrial Emission Directive 2010/75/EU

3. Illy, F., Borders, H., Joshi, M., Zuchelli, P., & Jurcik, B. (1998), "Processes for heat recovery and energy savings in oxy-fired glass furnaces: a technological survey", International Glass Journal, Vol. 96, pp. 65-72.

4. Joumani, Y., Leroux, B., Contino, A., Douxchamps, O., & Behen, J. (2010), "Oxygen and natural gas preheating for Oxy-float glass", Glass International, July/August, pp. 32-34.

5. Joumani, Y., Tsiava, R., Leroux, B., Contino, A., & Douxchamps, O. (2010), "ALGLASS heat recovery: an advanced oxy-combustion technology with heat recovery for glass furnaces makes sustainable performances", Verre, Vol. 16, No. 5, pp. 19-24.

Modeling

INTELLIGENT FURNACE DESIGN & CONTROL TO INCREASE OVERALL GLASS FURNACE EFFICIENCY

Erik Muijsenberg
Glass Service Inc.
Vsetin, Czech Republic

ABSTRACT

In the past decade, advanced multiple input/multiple output (MIMO) process control systems have found their way into the glass industry. Today over 115 glass furnaces and forehearths have been equipped with a supervisory control system. Daily regulation of fossil fuel firing and electric energy supply to stabilize temperatures is no longer in the hands of the operator, but fully taken over automatically by the ES III™ Model Based Predictive (MPC) controller, which provides consistent process control, 24 hours per day, focused to operate the entire glass production process in the most efficient way. Glass Service has developed the concept to design a more intelligent furnace concept that can use the benefits of a more intelligent control that can use more sensors (outputs) and more control variables (inputs). It offers a complete matrix control solution.

INTELLIGENT FURNACE DESIGN

The most common furnace used in the glass industry today is the End Fired furnace. It is popular because of the high thermal efficiency and lower investment costs. Part of this can be explained by the more compact regenerators and less burners and smaller batch chargers coming from the end firing concept, than coming from the (older) cross firing concept. However one disadvantage about the end-fired furnace is that it is more difficult to adjust the flame and so to create a hotspot by the top heating of the glass. Therefore some amount of electric boosting makes control of the glass melting easier, and at the same time it provides the furnace with a higher degree of operating flexibility. Further some electric boosting is almost a must when melting darker glasses. But, deciding to install boosting of let's say 800 kW on an optimal location can be done optimal or sub optimal.

The use of Computational Fluid Dynamics (CFD) simulation models can be used to calculate the model. As an example, we take a typical end-fired furnace and we study four (4) variations with different locations of the Electric boosting electrodes. Variation 1. Barrier Boost, Variation 2. Batch Melt Boost, Variation 3. Sidewall Boost, Variation 4. Centerline Boost. The models can calculate the effects of the different locations of the glass temperatures and glass convection loops. Then we trace massless particles within the glass melt to judge which variation gives the best melting conditions. In summary, the longest minimum residence time and the highest melting index should give the best melting performance.

Figure 1shows the four (4) design variations including the particle trace paths of some of the shortest residence times. The Left Top barrier boost variation effectively yields the most trace lines. The Right Top side boost warms the sidewalls very well, but leaves the shortest path only in the center. The Left Bottom Melter boost affects the least area, and leaves the shortest path in middle open. The Right Bottom centerline boost, stops the center shortcut, but also propagates the side flow without creating a so called spiral effect to improve the melting. (This would require more power than 800 kW.)

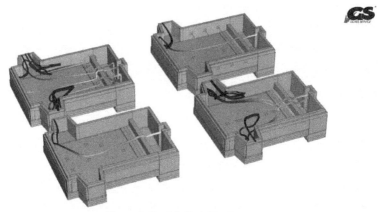

Figure 1. Particle Path Tracing

Table 1. Trace Results in Numbers

		Residence Time (minimum)	Melting Index ($\times 10^6$)
Melt	(Bottom Left)	3:43	1.01
Barrier	**(Top Left)**	**4:11**	**1.20**
Centerline	(Bottom Right)	3:10	0.95
Sidewall	(Top Right)	3:31	1.04

From the particle trace results we can conclude that for this furnace, glass composition and pull the Barrier Boost (Top Left illustration in Figure 1), just before the physical barrier, gives the best results (longest minimum residence time and highest melting index).

INTELLIGENT MODEL BASED PREDICTIVE CONTROL

Expert System, ESIII™, provides supervisory multi-variable model based predictive control of an operating furnace in time and space. This definition is not very understandable to people who are not familiar with advanced process control, thus let's try to explain it more detail.

The word "supervisory" means that the Expert System implementation becomes the operating control parameter at the primary control level: essentially the "supervisory" level controls all of the devices (PLCs, internal switches) as well as updating the control mechanisms

(PID loops, availability to change inputs manually). ES III™ can even use the primary control level elements as part of the advanced process control.

The next word, "multi-variable", means that multiple inputs can be handled at the same time to affect multiple outputs. This is very important for the supervisory control of the glass melting and conditioning processes. For example, in a float furnace, the total amount of gas is only one of the inputs that affects the crown temperatures. Other factors such as the batch feed rate and batch composition are important as well. Even the ambient air temperature, which impacts the combustion air preheat temperatures can impact the crown temperatures. So it can be concluded that crown temperatures are affected not only by one variable (total gas), but by many more. On the other hand, there are usually several crown and bottom temperatures with different levels of importance for the process control. And we have not discussed yet the effects of glass level, canal temperatures, oxy-boosting, etc. All of them affect (or are affected by) the process control too. Overall, the process control is a complex problem whereby multiple inputs and outputs interact strongly together at various levels of importance. When having experience with the Expert System the customer can decide to design a more intelligent furnace that makes use of more intelligent sensors and also more inputs to stabilize the production.

An important portion of the Expert System definition is the word "model" (model based predictive control). It refers to the control techniques that are mostly used. Model based control uses information about the process variable's history and utilizes a mathematical approximation of the process behavior: the "model". Model knowledge is important for process control and the predicted future behavior. At each moment, the next control action is planned as well as the estimated future behavior of temperatures and other controlled variables. The models can be derived in four (4) ways:

1. From a historical database.
2. From careful dynamic step response testing on the real process.
3. From step response tests on a CFD model (also called white models).
4. By a combination of the above or by estimating models based on the behavior of the process.

On top of this there are properties not mentioned in the definition above that are implemented in ES III™ too. For the architecture of the software see Figure 2.

Every system you are going to use for continuous glass production has to be safe. At first, it is necessary to return automatically to the basic control level in case the communication between systems (ES III™ and DCS) fails. Next, ES III™ contains fault detection logic that refuses to react on variable values that are out of specified intervals or changes abruptly (typical for thermocouple failure). An off-line simulator can be included into the safety tools too. Questions to be asked would include examples like: Would you like to know how the process will react with different process control settings? Or even would you like to try it but you don't want to violate your valuable glass production? The off-line simulator offers you the ability to experiment with these settings off-line, in simulation mode without affecting the process. Thus you can look for best-fitting solution that fully matches your ideas about precise process control while you are producing high-quality glass. After verification of these new process changes, the final solution can be applied as settings on the real process control. You also can store these settings through the Process Settings Manager. This manager contains recipes for various glass production types so these settings are always ready to be applied when changes are made to the process – it eliminates potential mistakes that can be caused by hand made inputs. Flexibility will come from the entire package of algorithms especially when devoted to process control. It can happen that some variables affecting process behavior are not predictable – in other words, you can estimate the future behavior only with limited reliability. As a consequence, the model based predictive controller (MPC) cannot consider these variables as known inputs. So it is necessary to find out alternative ways how to get such variables back to the game. Except for MPC controllers there are fuzzy logic controllers and classic Rules Based systems available for use. These tools are very powerful and can work either separately or together. It gives the process control designer a bigger chance to choose the strategy that fits optimally.

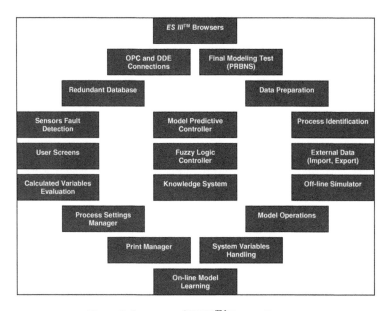

Figure 2. Summary of ES III™ System Features

Beside all of the properties mentioned above, there are several tools that belong more to the overall plant operational information. Among these tools can be included an internal database (which gathers and stores data in the requested form and timing), user screens (serves the purpose of visualization of all available data including data from the database) and reports (either printed at specified time of the day or stored as a PDF document), etc.

ES III™ is the trademark of the advanced process control system developed by Glass Service, Inc., and is especially designed for applications in the glass production industry. The glass melting process is characterized by slow process changes in the melting tank with extremely long reaction times, but on the other hand short timescales there are more rapid changes in the combustion process changes. Typically, glass resides 24 hours or longer in the melting-end and any change in the furnace heat input can easily take 4 till 8 hours before the impact on the glass melt throat or bottom temperature is observed. This makes it difficult for operators keep a high level of furnace stability in their limited 8 hour shifts, and also conventional automatic controllers without process knowledge or predictive response models are not suitable to handle such long term process changes and slow response (e.g. change of melt temperatures by changing fuel input) either.

The Model-based Predictive Controller (MPC) is the main control engine inside ES III™. Unique process models (identified from a 3-4 weeks historical data sample) are used. These models are so called black models and can be derived in different ways as described earlier, which informs the controller about the actual response of the furnace, including dead times between the heat input and change in the glass melt temperature, at different positions important for glass quality, over time. ES III™ continuously monitors the actual situation in the furnace and forehearth and predicts future trends based on the process models and recent control actions. If a new temperature set-point is required, the ES III™ system predicts the necessary changes in top fuel firing to achieve the new temperature in a fast and optimal way taking into consideration constraints (e.g. maximum temperatures or maximum acceptable fuel injection rates), just like a car navigation system optimizes the directions to a new destination. Despite a good process model, the glass melting process is unfortunately not 100% predictable because of disturbances that cannot be measured nor predicted, so (just like the weather forecast), hence a continuous adjustment of the long term prediction is inevitably necessary. For glass melting and conditioning, the prediction update of the process every couple of minutes is sufficient and the operator is informed by clear trend lines in the ES III™ graphical user interface (see figure 3).

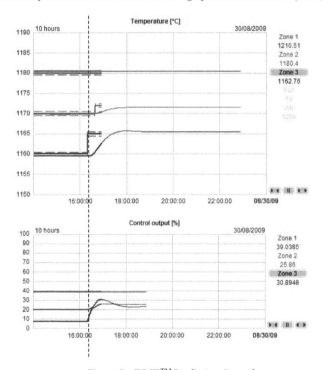

Figure 3 - ES III™ Predictive Control

In figure 3, the thin solid trend lines show the predictions of the temperatures (top) and planned firing changes (control output, bottom picture) in a forehearth with 3 conditioning zones. The dashed vertical line denotes the actual time. The solid thick lines show historical behavior and actions and the dashed lines are the zones of the controlled variable (which can also be called a set point within a zone).

CONTROL AND STABILITY

The primary objective of the ES III™ controller is to control the temperatures of the entire glass melting process and improve their stability, including the glass melt level and the final melt temperatures delivered into the forming process (e.g. gob temperature or glass outflow), firing and cooling conditions, etc. More stable production conditions obviously yield better conditions for glass forming, thus increasing the production line yield and minimizing the amount of rejects. Customers even report reductions of common hot repairs and thus a positive impact on the furnace lifetime, however it is not easy to prove that this is directly related to only the better control.

In order to take over the entire temperature control of a glass furnace, the advanced controller should be able to control multiple temperatures at the same time. Typically, the furnace crown temperature and temperature profile should be controlled to prevent crown overheating (or to prevent too low temperatures where silica reacts with sodium vapors) and to obtain the minimum required bottom glass temperature to melt the glass with the best glass quality. Usually, there is more than a single burner in the furnace, and usually additional electric boosting is directly applied into the glass melt. All these heat sources should operate together in the most efficient and stable condition. The ES III™ MPC controller has been designed to handle such Multiple Input Multiple Output (MIMO) process changes that takes into account all possible correlations between the various inputs (firing, boosting, valves, cullet ratio, pull) and control outputs (temperatures, glass melt level, and sometimes pressure).

HOW TO SAVE ENERGY

Improving the stability of the melting process is the first step towards saving energy. A continuous monitoring of the process helps to prevent glass temperatures to slowly drift and excessive reactions of the operators in firing may then be necessary. The ES III™ system watches the furnace the entire day and collects all relevant furnace operating data to update the predictions of the future trend lines. Only smooth changes are applied in the firing and furnace setup, unless a major step change is required (e.g. at a sudden job change). It is important for energy efficiency that throat temperatures are not higher than really necessary for a certain glass product, as higher temperatures result in extra energy consumption in the glass furnaces and requires additional cooling in the forehearths, working-end or feeders. However, on a long term basis, the expected savings in terms of heat input often turn out to be marginal typically: 1-4 %, though most traditional glass furnaces are operated by experienced operators that are able to control the process without excesses most of the time. This situation can be different in completely new plants as new operating people do not have so much experience and need to learn from their mistakes. Enhanced stability of the glass melting and conditioning process is often most profitable for the subsequent glass forming, hence more and better quality of final products and higher pack-to-melt levels.

The figure below shows how temperature control and stability can really be used to save energy. Most glass furnaces are operated at an average glass temperature that is hot enough to prevent fining problems and seeds and blisters in the final product. The more the glass temperature varies, the hotter the average furnace glass melt temperature is chosen to prevent glass quality issues. With a controlled and stable glass temperature, the furnace manager is able to explore the limits for his glass production. By lowering the set-point of the operating temperature the furnace manager can save a considerable amount of energy, up to 3% or sometimes even more. But this depends also on how the customer uses the tool most effectively.

Figure 4 shows energy savings by temperature control and stability. The step down in the set-point of the glass temperature saves directly on average gas consumption by using the reduction in temperature variation to explore the actual limits of the melting process and to keep the desired glass quality. The vertical scale on the top picture shows temperature in °C and on the bottom picture shows the gas consumption in m3/hr. (Note: For customer confidentiality the real levels are not shown.)

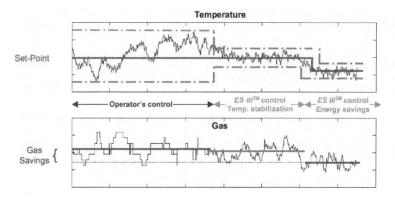

Figure 4. Energy Savings

As a third example of how to save energy using an advanced control system, let's comment on a conventional forehearth control. Usually each conditioning zone is controlled by a individual PID controller with the temperature set-point selected by an operator. As the zones are controlled individually, the firing in each zone can be completely different. Often the heating in one zone is followed by cooling in the next zone and again strong heating in the next. A smooth reduction in the forehearth temperature with more uniform heating or cooling is considered more ideal and this can be accomplished with a greater degree of accuracy with the advanced ES III™ MIMO controller that is able to handle all of the zones of the entire forehearth at once. Cooling valves are only opened after firing is reduced to the minimum and, oppositely, are properly closed before heating starts. The MIMO controller can also help to make the glass melt temperatures close to the spout (e.g. at the 9 grid if available) more homogeneous, but practical experience has shown us that keeping the glass melt temperature of the 9 grid bottom measurement constant is more important than small homogeneity variations. Also the model-based control can help to change these temperatures faster than operators can do for a job change (typically twice as fast).

The last example mentioned here applies to fossil fuel fired glass furnaces that are equipped with an additional electric boosting system. The electric boosting systems are usually very efficient as the heat is supplied directly to the glass. However, the high price of electric energy often forces the furnace manager to find an optimal balance between energy efficiency and cost efficiency. Using the ES III™ Minimal Cost Strategy, the MPC controller takes into account the prices of the various heat sources as well as their efficiency to control the furnace temperature. So the required glass temperature is obtained at minimum costs. The same strategy can even be used to find the optimal balance with two (2) or more electric boosting systems. It's almost impossible to find such a balance without a supervisory control system.

The ES III™ advanced control system is a tool that helps the operator in their daily operation of the furnace, as well as the furnace manager being able to reach his targets for producing glass of the highest quality with minimum energy and costs.

Typical energy savings at most of our customers range from 2-4%. Actual results depend also on how customer uses the options that ES III™ are giving them. It means the typical Return of Investment (ROI) of an installation is matter of months, far less than 1 year.

MAKING THE FURNACE MORE INTELLIGENT

Humans are operating with many sensors that feeds our brain; we have eyes to see, noses to smell, ears to hear, fingers to touch, tongues to taste, on the skin and internally we also have temperature sensors. All of these sensors feed our brain and our brain (hypothalamus) decides what to do when we are overheating or cooling down. A typical Glass furnace has a limited amount of sensors and some of them indicate a reaction when it is already too late and more difficult to stop the trend. Such an example is the change of the gas composition that effects the heating value of the gas that in turn will affect the furnace temperatures, but can be detected earlier by measuring the combustion enthalpy or even better analyze the actual gas composition.

If the customer already has experience with controlling the furnace with a supervisory control system like the Expert System III, then it makes sense to add more sensors. If the control is done by the operator, the use will be limited, as how the operator will interpret the effect on the furnace if Hydrogen increased by 4% in the natural gas instead of Methane?

Figure 5 shows possible sensors and loops that are available today and can be added to make a fully automatic control system possible.

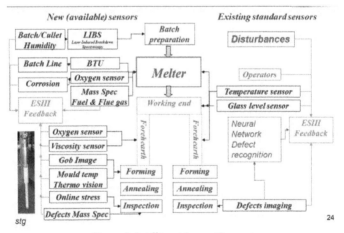

Figure 5. Intelligent Sensor Concept

One such example of a new sensor is the Batch Vision sensor where we digitize the signal of the furnace camera to calculate the exact position and size of the batch blanket cover.

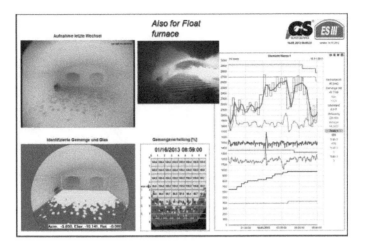

Figure 6. Batch Vision inside an End-Fired furnace

CONCLUSIONS

Most planes fly today 98% of their flight with an autopilot. In 2025, most cars will be able to drive themselves on our highways. In the future, most of glass furnaces will perhaps be controlled full automatic most of the time in a closed loop control mode.

REFERENCES

[1] Josef Müller, Robert Bódi, Josef Chmelař: How to Make Glass Furnace Control Easier: Advanced Optimal Control by Expert System *ES III^{TM}*, Proceedings of the VIII. International seminar on mathematical simulation in glass melting, Velké Karlovice 2005, p. 200

[2] Josef Müller, Josef Chmelař, Robert Bódi, Erik Muysenberg: Aspects of Glass Production Optimal Control, Proceedings of the VII. International seminar on mathematical simulation in glass melting, Velké Karlovice 2003, p. 157

[3] Josef Müller, Robert Bódi, František Matuštík: Expert System ES III - A System for Optimal Glass Furnace Control, Glass - Monthly Journal (Vol. 81, No. 9, October 2004)

[4] Josef Chmelař, Erik Muysenberg, Robert Bódi: Optimizing Glass Production with Expert System ES-II, Proceedings of the VI. International seminar on mathematical simulation in glass melting, Velké Karlovice 2001, p. 101

[5] Erik Muysenberg, Josef Chmelař, Robert Bódi: Supervisory Advanced Control of Glass Melters by GS Expert System ES-II, Proceedings of the V. International seminar on mathematical simulation in glass melting, Horní Bečva 1999, p. 162

[6] Peter Mikulecký, Robert Bódi, Josef Chmelař: Towards Total Glass Quality Management, Proceedings of the IV. International seminar on mathematical simulation in glass melting, Horní Bečva 1997, p. 172

GLASS MELT QUALITY OPTIMIZATION BY CFD SIMULATIONS AND LABORATORY EXPERIMENTS

A.F.J.A. Habraken, A.M. Lankhorst, O.S. Verheijen, and M. Rongen
CelSian Glass & Solar B.V., Eindhoven
The Netherlands

ABSTRACT
    The quality of a glass melting process is determined by the melting-in performance, the sand grain dissolution and the removal of gas bubbles by fining. For good quality it is essential that each trajectory/path starting from batch charging to the throat or waist of the tank shows complete melting and fining. Each part of the glass melt should be exposed to temperatures such that fining can occur. The fining onset temperature and melting performance for an industrial glass tank depends on the batch composition, sand grain size, type and amount of fining agent, furnace atmosphere and the redox number of the batch. Laboratory experiments are carried out to determine the most important parameters for industrial batches. The results of these laboratory experiments are used as input to the CFD (Computational Fluid Dynamics) model GTM-X, which then provides the temperature and flow fields in the industrial melting tank, along with the calculation of the individual glass melt trajectories. This combination of experiments and CFD simulation proves to be a very powerful tool for determining the fining and melting performance of industrial furnaces. Special attention will be paid to the location of the fining zone, with respect to the convection flows in the melt.

1. INTRODUCTION
    The glass melting process can be divided in different subsequent process steps [1]. In an ideal furnace (with respect to product quality), these process steps would be separated in single zones assuring that each process step is optimally performed. Most glass melting furnaces these days are based on a different concept in which the process steps interfere with each other. Figure 1 provides an overview for such furnaces. Raw materials fuse under the impact of high temperatures, and fresh melt moves towards the fining area where bubbles and gasses are removed from the glass melt. The remaining melt, containing small bubbles only, proceeds towards the refining area where these bubbles dissolve in the glass melt. For homogeneity it is important that the glass melt is well mixed. This mixing occurs during each of the process steps, and results in the exchange of glass melt between the process steps. In an industrial float furnace it means that a significant part of the good quality glass melt from the refining step moves back into the furnace and mixes again with freshly molten glass melt.

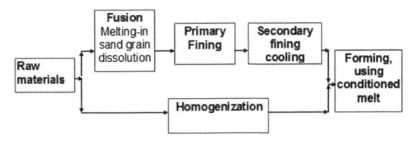

Figure 1. Process steps in conventional glass melting furnaces.

In current conventional float furnaces, it is unavoidable that good quality glass melt mixes with freshly molten raw materials. Finding optimum process settings and optimal furnace design such that a minimum of mixing is required while maintaining to match product specifications is a very difficult task. Computer modeling is a very helpful tool to control the mixing in the furnaces.

Although computer modeling is a tool that often is applied in the glass industry, it remains difficult to have a waterproof prediction on product quality. The techniques used to predict product quality are often based on indices that provide a relative prediction while glass producers are dealing with absolute numbers of glass faults prescribed by their customers. Quantitative approaches are available, but are complex and expensive. The approach given in this paper is practical and has proven itself in industrial projects for the glass industry.

2. GLASS MELT QUALITY PREDICTION

The quality of a glass melting process is determined by the melting-in performance, the sand grain dissolution and the removal of gas bubbles by fining [2]. For a good quality it is essential that each trajectory/path starting from batch charging to the throat or waist of the tank shows complete melting and fining. Each trajectory should be exposed during a sufficiently long time to a temperature above the fining onset temperature. The fining onset temperature and melting performance for an industrial glass depends on the batch composition, sand grain size, type and amount of fining agent, furnace atmosphere and the redox number of the batch. Calculating all chemical reactions in the glass melt would provide information on batch melting behavior and the release of gasses in the glass melt. For the chemical reactions that predict fining and refining, the transport of species in the glass melt and chemical equilibrium reactions can be calculated. Equation [1] shows the equation for this calculation.

$$G(\mathbf{n}) = \sum_{j=1}^{Nf} \mu_j^0 n_j + \sum_{j=Nf+1}^{Nf+Ng} \mu_j^0 n_j + \sum_{j=1}^{Nf} RT P_j \log\left(\frac{P_j n_j}{\mathbf{P}^T \mathbf{n}_m}\right) + \sum_{j=1}^{Nf+Ng} RT \log\left(\frac{n_j}{\mathbf{1}^T \mathbf{n}_g}\right) \qquad [1]$$

Results of the chemical model are used in combination with bubble trace analysis, such that an amount of bubbles in the product is predicted. Although this method can be very accurate and elegant, the thermodynamic input parameters for this model are difficult to determine and therefore expensive. Choosing for a practical and reliable method, it was decided to couple the bubble analysis in the computer model to existing laboratory experiments which are used to observe and study melting behavior by video observation and analysis of released gasses (HTMOS system at CelSian). Figure 2 shows the three main process steps that are monitored, and from which results are used as input for the computer model.

From monitoring the batch melting, input parameters are determined for the batch model and the sand grain dissolution model for a specific glass melt at industrial conditions (equal environment is simulated). Returning to the topic of fining and refining, the complex chemical calculations mentioned previously are now replaced by the output of the HTMOS system. Using image processing software (Figure 3), the size and growth of bubbles can be monitored and stored during each phase of the experiment.

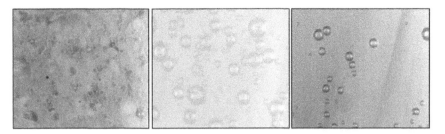

Figure 2. Melting, fining, and refining behavior monitored in the high temperature observation system.

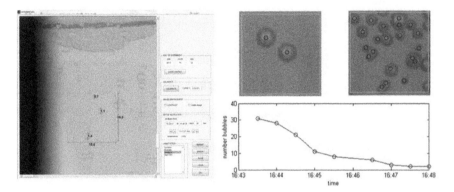

Figure 3. Image processing software for bubble growth investigation.

Once the data from the melting experiments is incorporated in the computer model, the quality analysis is executed in two steps:

1. Sand grain dissolution trace, following sand grains from batch charger to the moment of complete fusion;
2. Bubble trace analysis, following bubbles from the moment sand grains are dissolved and bubbles start to grow, to the end position of the bubbles.

Amount of bubbles in the product is used as an indicator for product quality. Excellent correlations have been found between percentage of bubbles predicted by this approach and glass defects (seeds and blisters) measured under varying conditions.

3. DEMONSTRATION CASE

In this part, a demonstration furnace [3] is selected and parameter studies are discussed that investigate glass product quality for those cases where the main glass flow patterns are changed. The chosen furnace configuration is an oxyfuel fired float furnace (Figure 4) producing flint glass (800 ppm Fe_2O_3).

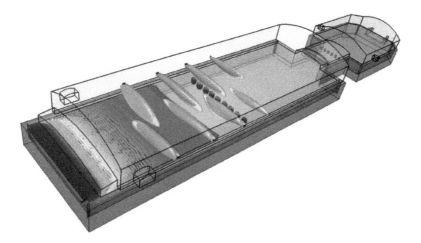

Figure 4. The oxy-fuel float furnace under consideration.

The main characteristics of the furnace are summarized in Table 1.

Table 1. Characteristics of the oxy-fuel furnace.

Fuel type	Oxy/gas-fired
Burners	Flat flame & pipe-in-pipe
Pull	150 ton/day
Firing rate	~ 11 MW (~ 1250 m_n^3/hr)
Melter	Length 23 m, width 7 m
Waist	Length 4.7 m, width 2.8 m
Working-end	Length 7.5 m, width 5.8 m
Melting energy	7.76×10^5 J/kg
Cullet fraction	25%

For the base case study, the results of the sand grain dissolution are shown in Figure 5. The model shows that most of the batch material fuses in the direct vicinity of the batch tip. This observation is in line with statements in literature [1]. Using the end position of the sand grains, mostly near the batch tip, a bubble trace is started. Bubbles start with a range in bubble diameters as is observed in the HTMOS system. One time-snap from the bubble traces is shown in Figure 6. Most bubbles escape from the glass melt in the hot area between the batch tip and the bubbler row. A few small bubbles stick to the sidewall and pass the bubbler row. These bubbles partly dissolve in the working end and partly end up at the lip.

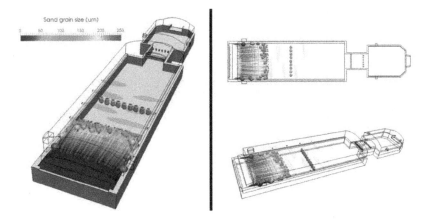

Figure 5. Sand grain dissolution model, showing that melting is most effective at the tip of the batch blanket.

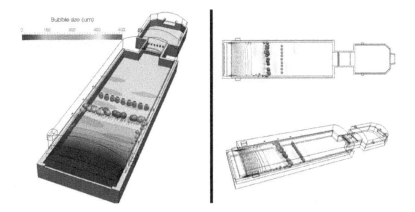

Figure 6. Bubble trace model, showing that fining is optimal in the large hot fining area between the batch tip and the bubbler row.

For a float furnace as described in this paper, glass melt flow patterns can be influenced by changing A) furnace geometry, B) firing profile, C) cooling in the working end and waist, D) bubbling amount and location, E) electrical boosting amount and location, and F) stirring.

Investigating the impact of the hotspot position with respect to the spring zone, the results of the following parameter studies are discussed here:

1) Var 1: Change firing distribution: move the maximum firing from the batch tip downstream towards the bubbling row;

2) Var 2: Move the bubbling row 2 meters upstream towards the batch blanket, near the hotspot position in the base case;

3) Var 3: Move the bubbler row 2 meters downstream.

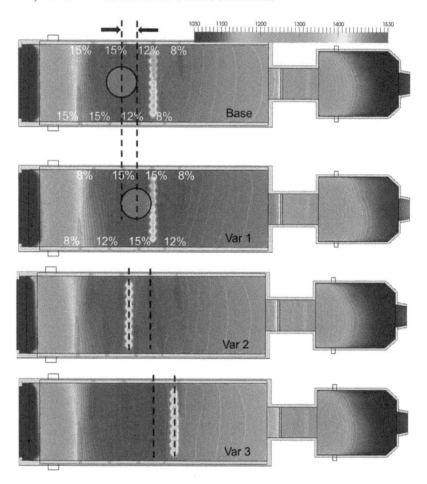

Figure 7. Bubble trace model, showing that fining is optimal in the large hot fining area between the batch tip and the bubbler row.

Temperatures on the glass melt surface are shown in Figure 7 for the base case and the variation studies. The difference between the base case and Variation 2 is the largest. It seems that the hotspot surface temperature decreases as an effect of the colder bottom glass pushed by the bubblers to the melt surface.

Fining is seriously affected by the lower melt hotspot temperature, as well as by the reduced fining area between the batch tip and the bubbler row. Bubble tracing reveals that a significant amount of bubbles will end up in the product in case of Variation 2 (see Figure 8). Plotting the entire distribution of initial bubble diameters versus bubbles in product for these model studies produces the graph in Figure 9.

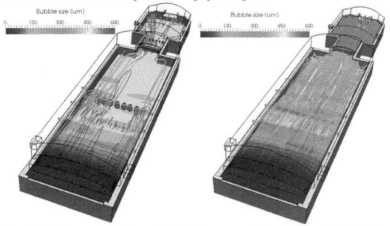

Figure 8. Bubble trace analysis studies for the base case (left) and the variation (right) in which the bubbler row was moved 2 meters upstream towards the original hotspot position.

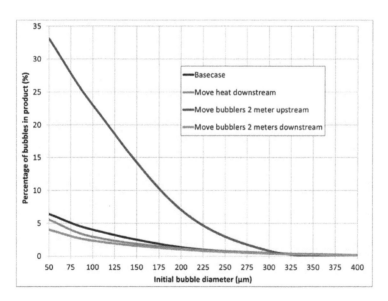

Figure 9. Percentage of bubbles in product versus initial bubble size

Large bubbles escape easy from the glass melt. Figure 9 shows that for the initial bubbles larger than 300 micron, no bubbles end in the product. For small bubbles, the optimal configuration is Variation 3, in which the bubblers are positioned 2 meters downstream.

4. CONCLUSIONS

Application of laboratory experiments in CFD modeling with GTM-X:

- Is a good alternative to detailed redox calculations to find the parameters describing glass melt quality in computer modeling;
- Results in excellent prediction of the batch blanket position, and its effect on the melt flow:
    o Density differences between raw materials and glass melt is an important driving force for recirculation flows in the melter;
- Provides reliable input for sand grain dissolution and bubble trace computer models;
- Provides additional information on the melting process (like presence/amount foam) what is often difficult to measure in the actual furnace;
- Also provides valuable information on new melting processes/glass types, which are not already in production.

Quality analysis techniques:

- Flow patterns and melt temperature profiles do not reveal furnace melting quality behavior:
    o The 3-dimensional fields are too complex to understand without analysis tools;
- Drawing conclusions that are based on quality indices from inert particle tracing, provides very limited and often inaccurate information on the melting, mixing, fining and refining behavior of the furnace.
- Performing sand grain dissolution and bubble tracing analysis provides a much more complete view on furnace behavior with respect to product quality;
- Starting from a validated base case, including the full analysis described above has proven to be a reliable and successful method for furnace optimization studies.

Effect of changing bubbler positions:

- When moving the bubbler position beneath the hotspot, the hotspot temperature will decrease resulting in poor fining behavior;
- The size of the hot fining area has a significant impact on the furnace's ability to remove bubbles. When increasing this area, by moving the bubbler row 2 meter downstream, away from the hotspot, the predicted furnace product quality is optimal;

These observations are valid for this furnace, but may differ for furnaces of other sizes, production, glass type, and configuration, as these parameters also influence the product flow quality and are presently not investigated in this study. Using these analysis techniques supports the customer in achieving his requirements and to find the optimum between cost price and necessary product quality.

REFERENCES

[1] R. Beerkens, Analysis of elementary process steps in industrial glass melting tanks – some ideas on innovations in industrial glass melting, Ceramics – Silikaty 52 (4) 206-217, 2008.

[2] A. M. Lankhorst, Advances in Fusion & Processing of Glass, 25 – 30 May 2014, Aachen, Germany.

[3] A. M. Lankhorst, L. Thielen, A.F.J.A. Habraken, Proper modeling of radiative heat transfer in ultra-clear glass melts, 73rd Glass Problems Conference, Oct 1 – 3, 2012, Cincinnati, OH, USA.

Forming

FOREHEARTH HEATING

Alan Stephens, Clive Morgan, Stephen Sherlock
Fives Stein Ltd
4A Churchward, Didcot, Oxon, OX11 7HB, UK

ABSTACT

For many years now incremental improvements have been made to forehearth heating systems that have made them more efficient, more controllable and safer. Fundamentally; however, they are still mostly K-type firings systems, which were introduced in the 1920s. The system feeds a mixture of gas and air to rows of small pencil burners arranged down either side of the forehearth channels. The efficiency and distribution of the heat inside the forehearth is dictated and limited by this arrangement and further improvements are limited. This paper discusses the development and trials of an alternative combustion system that uses fewer burners, distributes the heat better and more efficiently and provides the opportunity for direct heat recovery, which cannot be achieved using pre-mix systems. The combustion controls for this new system offers the same accuracy of control as the best existing systems but with the potential for an even higher turn-down ratio, improved safety, and direct heat recovery. The burners provide better heat distribution and more efficient combustion, resulting in improved side heating, compared to the traditional K-type firing systems. Real fuel savings should result from the wider turn-down ratio and improved combustion efficiency, whilst the improved heat distribution will allow improvements in the thermal homogeneity of the glass leaving the forehearth.

INTRODUCTION

The control of the glass temperature in the forehearth has become critically important to the successful operation of today's high speed production lines. The ability to attain this level of control at high throughputs has been achieved by major improvements in structural design, especially relating to cooling, improved components and microprocessor based control systems. The forehearth heating system too has improved, with better flow control and mixing devices. But, in the majority of cases, the final delivery of heat is still through the type and arrangement of burners first used in the 1920s (Figure 1). The materials may have improved but the fundamental principle of heat delivery has not.

The standard "K type" firing system, which is the basis of most forehearth gas heating systems, is an arrangement of small pre-mix "pencil" burners at 4½" centres down both sides of the forehearth channel. This helps control the heat losses from the slower sidewall glass flows, but the heat release is still not ideal. It has been know for some time that this type of flame attains its highest temperature well forward of its root, which puts the main heat release beyond the edge of the channel block. The narrow spacing between burners is therefore necessary to minimise this effect and provide, as near as possible, an even longitudinal distribution of heat.

DEVELOPMENT

For many years we have been working to improve and simplify the combustion system and have:

- Increased the firing range by the use of better air-gas mixing (from 3:1 to 6:1+)
- Increased the controllability by the use of variable speed fans
- Improved the safety by using self-contained zone systems.

However, none of these has overcome the fundamental issue of heat release inside the forehearth. To address this limitation we needed to look at the combustion process itself.

Figure 1. K-type forehearth firing system

We have worked on oxy-gas firing, which can achieve the required sidewall heat release profile but, for mainstream cooling forehearths, this is not a cost effective solution. We have also investigated oxygen enrichment. Again this would improve the sidewall heat release. The ideal way for this to be done would be to produce one stream of oxygen enriched air and feed it to each combustion zone, which could then remain fundamentally unchanged. The only added complexity would then be one oxygen enrichment unit for the combustion air. Unfortunately, US gas safety regulations prohibit the use of oxygen enriched air in air-gas mixing units so any oxygen enrichment would have to be done after the normal air-gas mixing. This then makes this arrangement far more complex and far less attractive as a way of improving combustion.

In looking to improve air gas mixing, we have developed a tee-mixer system using a proportioning regulator. This uses the air flow through an orifice plate to regulate the flow of gas to the mixer and offers better air-gas ratio stability over the full firing range than can be achieved with a venturi mixer. In this system the tee-mixer is not a fundamental part of the air-gas ratio control, but only a means of mixing the air and gas. It was therefore a simple step to move to nozzle mix burners. This was actually a different development for a high temperature application, where much larger burners were required, but this offered us another tool we could use to improve general forehearth combustion. Despite these improvements; however, what we needed was a slower flame over a wider flame front.

We began looking at various ways of slowing and widening the flame front to try and achieve the better sidewall heat release we desired. Many different arrangements of burners were modelled and several rough prototype burner blocks were created and tested. The result of which was a burner block that seemed to deliver what we required (Figure 2).

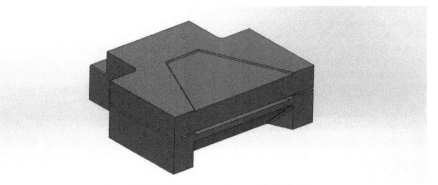

Figure 2. Model of the new burner block design.

The internal shape of the burner block is important to the operation of the burner, but posed some manufacturing problems. This was too complex to be easily produced as a one piece block, but it is important that the internal chamber is not compromised. The solution was a two piece construction that is bonded into a single block before it is fired. The burner blocks provide the direct support for the forehearth roof blocks, which puts them under mechanical stress. These new blocks would also be under higher thermal stress so it was important to ensure that the design compensated for that and placed the mechanical loading only on the outer, solid sections of the burner block.

The refractory material of the block had to be suitable for the casting and bonding process but also had to be thermally shock resistant and suitable for most glass applications. Our normal burner block material, a bonded alumina-silicate, was found to be unsuitable but a high thermal shock resistant mullite did prove to be suitable.

PROTOYPE TRIALS

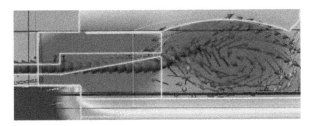

Figure 3. CFD modelling results showing temperature field in the forehearth cross-section.

The CFD modelling results, Figure 3, shows that the flame is lifted onto the underside of the roof successfully concentrating heat onto the sidewall of the forehearth. The wider flame front is also slower moving and thus appears to give up more of its heat in the critical region of the forehearth. In order to test this properly, a full scale zone superstructure was constructed. The test rig was first built with firing using conventional burners to give a benchmark to compare the new burner system against. For each test run the rig was heated up to normal forehearth operating temperature and then allowed to soak. Once temperatures were stable, a set of data was collected and then the temperature was increased. This process was repeated for a series of increasing temperature steps and then for a corresponding decreasing temperature steps. Once the test was complete the rig was cooled down and rebuilt with the new burner blocks and the process repeated. The results showed that the new burner did provide more heat to the sidewall and actually achieved it using approximately 10% less gas. A plausible explanation for this fuel saving is that the slower flame front allows it to give up more of its energy in the critical side region of the forehearth. Further tests have confirmed the improvements and shown:

- Excellent flame stability
- A wide turndown (8:1+), allowing greater flexibility
- Full clean combustion leading to a slower flow of combustion products, which causes less disturbance to the centre-line cooling.
- On our test rig at least, the use of the nozzle mix burner system has also proved to be much simpler to maintain than the conventional pre-mix system.
- The nozzle mix system is also inherently safer than a pre-mix system since no combustible mixture is being moved through the production environment.

A full forehearth firing system using the now patented Prium® Flat Flame burner blocks and nozzle mix combustion system has now been installed onto a forehearth in Poland, which is expected to go into production 19th October after which more information about its performance will be made available.

CONCLUSION

An alternative forehearth combustion system has been successfully developed that distributes heat better and more efficiently, and provides the opportunity for direct heat recovery. In tests the burner has attained its objective of providing heat closer to the forehearth channel sidewalls and it has achieved this using 10% less gas than the traditional K-type forehearth firing systems. The combination of the Prium® Flat Flame burner blocks and nozzle mix combustion system has also shown itself to be more flexible with the potential for further improvements in forehearth operation. A full forehearth firing system using the now patented Prium® Flat Flame burner blocks and nozzle mix combustion system has now been installed onto a forehearth which is expected to go into production on the 19th October. More information about its performance will be made available following the system start-up.

IMPROVEMENTS TO EMHART GLASS VERTIFLOW MOLD COOLING APPLICATIONS IN GLASS CONTAINER PRODUCTION

Pierre S Lankeu Ngankeu
Bucher Emhart Glass
Windsor, CT USA

ABSTRACT

During the forming process molds are used to cool the glass while it is blown into its final shape. The heat removed from the glass is dependent on mold temperature. This makes mold cooling one of the most critical aspects of the glass container production. Air is often used as a cooling medium. For over 30 years, Emhart glass has been using the Vertiflow cooling system on the Blow (or mold) side. This system provides cooling air to the molds from the bottom plate mechanism. This system has always been limited to cooling only being available while the molds are closed. Some processes, namely those with smaller contact times on the blow side would benefit from additional cooling time. In addition, because the Vertiflow cooling air travels from the bottom to the top, it is difficult to achieve a good vertical temperature distribution for containers that require more cooling in the upper parts of the mold. In order to increase the flexibility of the Vertiflow cooling system on the blow side, Emhart Glass has developed Vertiflow Assist. This system complements the Vertiflow system by enabling cooling when the molds are closed. Vertiflow can be used to cool the molds while they are closed and Vertiflow assist is used while the mold is open. In this paper we will present the existing Vertiflow system along with the Vertiflow assist concept as it has been implemented into the AIS and NIS machines.

INTRODUCTION

The Vertiflow process has been used for over 30 years in the glass container production. At the time of its introduction, the standard cooling was done by blowing air in the back of the molds using fins. But, this method known as stack cooling (Figure1) was unable to capitalize on the improvements in mechanisms speeds. Vertiflow cooling (Figure 2) through holes in the molds coming from the bottom plate allowed for a significant speed increase compared to the previous system. However, Vertiflow came with its own sets of limitations. The most crucial was the fact that it could only be available when the molds were closed as it was fed from the bottom plate mechanisms. In addition, the bottom part of the mold tended to be overcooled. As machine production speeds have increased, there has been a push to improve the Vertiflow mold cooling process. Recently we have developed a solution for the parallel motion machines (AIS, NIS). This new system called Vertiflow assist will work in conjunction with the existing Vertiflow cooling to achieve higher production speeds and better container quality.

Stack Cooling

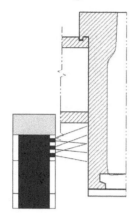

- Advantages
  - Versatile cooling
  - no valve = no downtime

- Drawbacks
  - Inefficient
  - Poor Control / Target areas finned to give greater cooling

Figure 1. Stack cooling

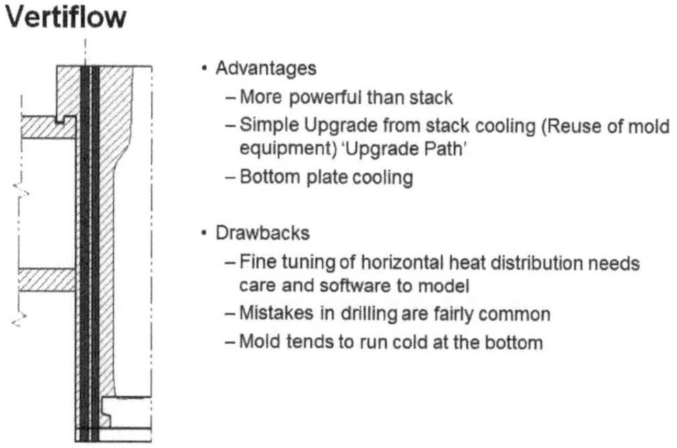

Vertiflow

- Advantages
  - More powerful than stack
  - Simple Upgrade from stack cooling (Reuse of mold equipment) 'Upgrade Path'
  - Bottom plate cooling

- Drawbacks
  - Fine tuning of horizontal heat distribution needs care and software to model
  - Mistakes in drilling are fairly common
  - Mold tends to run cold at the bottom

Figure 2. Vertiflow cooling

MOLD COOLING OVERVIEW

During the final blow process in the glass container forming process, the parison is blown into its final shape in a blow mold. At this time, the mold is used to remove as much heat as possible from the glass. This is done primarily through conduction as the glass is in intimate contact with the mold. The glass heats up the mold which is in turn cooled by convection, blowing cold air (typically between 15°C and 55°C) in holes through the mold (Vertiflow) or on fins in the back of the mold (stack cooling). The glass contact and the mold cooling time determine how much heat can be removed from the glass and how quickly the container can be taken out of the blow mold. Moreover, the mold must not be too hot or it will create hot spots by sticking to the glass. On the other hand, it cannot be too cold as it will create cold spots and results in cracks on the glass surface. It is imperative to keep mold temperatures within a specific working range depending on the material (400°C to 600°C for Cast iron molds).

VERTIFLOW COOLING

The vertiflow cooling system consists of a pressure manifold mounted below the bottom plate mechanism. Air is fed through the holes in the bottom plate to holes in the mold and escapes on the top to atmosphere. Because the air comes from the bottom plate it is highly efficient at cooling the bottom of the mold. As the air goes up it is usually warmer and not as efficient in the upper areas of the mold (especially for taller containers). For taller containers significant differences in temperature have been observed between the heel and the shoulder of the mold. These differences can lead to containers being out of round (oval) and out of vertical (leaners). In addition, due to its location on the bottom plate mechanism, the vertiflow air can only be used when the mold halves are closed. This can reduce the mold cooling time significantly (notably for container with short contact time). This disadvantage can reduce the machine speed for some containers (Figure 3).

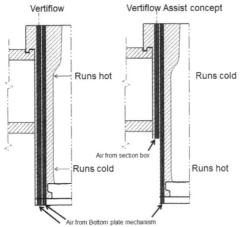

Figure3. Vertiflow vs. Vertiflow assist

VERTIFLOW ASSIST

In order to address the limitations of the vertiflow mold cooling system, we started a series of computer simulations coupled with experiments at our facility in Windsor, CT. The focus was put onto the existing parallel motion machines (AIS, NIS). Because of the differences in the design the vertiflow assist system would look significantly different between both machines. One of the challenges was that any new system would have to be compatible with existing equipment already in use by our customers to minimize cost. To this we had to make sure that the vertiflow assist system be able to integrate in the existing machine without too many changes. The vertiflow assist would have to be available when the mold halves are open. To that end we found a way to access air in the section boxes and reroute it for use in the Blowside. Initial computer simulations showed a potential 50°C to 100°C overall cooling effect compared to the existing vertiflow system. We continued with a series of test on a machine for both the AIS and the NIS.

VERTIFLOW ASSIST AIS

The AIS system uses the mold holder arm to supply air to the back of the molds (Figure4). The air is taken from the section box by a telescoping tube. A test trial was conducted on a Triple gob (TG) section at a customer facility in November 2014. The results showed a significant improvement in the vertical temperature distribution as well as the circumferential one (Figures 5-7).

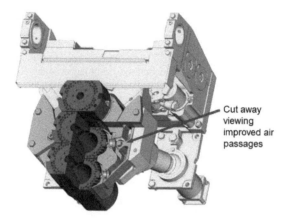

Figure 4. AIS Vertiflow Assist

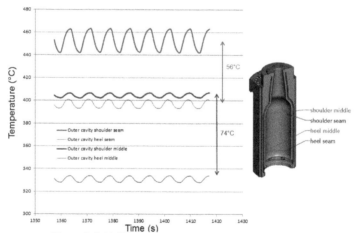

Figure 5. Initial Vertiflow vertical temperature difference

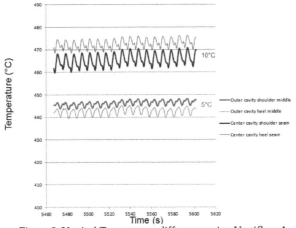

Figure 6. Vertical Temperature difference using Vertiflow Assist AIS

VF ON	VF OFF	Shoulder middle	Heel middle	Shoulder seam	Heel seam
Baseline	Baseline	402°C	325°C	448°C	352°C
10	175	346°C	303°C	372°C	336°C
30	175	354°C	318°C	380°C	351°C
50	175	369°C	343°C	408°C	378°C
70	175	385°C	362°C	414°C	398°C
90	175	394°C	388°C	437°C	429°C
110	175	422°C	413°C	456°C	454°C
130	175	452°C	427°C	469°C	476°C

Using VF assist we were able to reduce the Shoulder to
heel temperature difference by reducing the VF ON time

Figure 7. Using Vertiflow Assist to balance the vertical Temperature

VERTIFLOW ASSIST NIS

Unlike the AIS, the NIS system uses a plenum chamber inserted on the back of the mold. The plenum chamber is supplied from the section using two telescopic tubes. The first row of the mold gets air from the existing vertiflow system while the second row is fed by the vertiflow assist plenum chamber (Figure8). Tests were conducted at our facility and customer sites throughout the year 2013 on both TG and quad gob (QG) sections. The results showed an overall temperature reduction of 70°C compared to the existing vertiflow system (Figures 9-10).

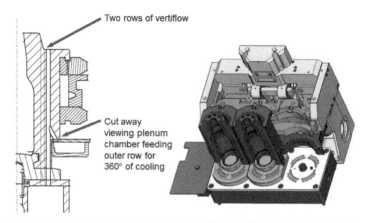

Two rows of vertiflow

Cut away
viewing plenum
chamber feeding
outer row for
360° of cooling

Figure 8. Vertiflow Assist NIS

Test 3	VF (°)	VF Assist (°)	Shoulder (°C)	Body (°C)	Heel (°C)	
Section 3	140	N/A	396.35	403.00	360.24	
Section 4	140	140	385.06	391.45	368.73	
			11.29	11.54	-8.49	ΔT
Test 4						
Section 3	140	N/A	399.50	405.94	363.87	
Section 4	140	280	342.10	349.09	333.33	
			57.40	56.85	30.54	ΔT
Test 5						
Section 3	140	N/A	398.32	404.51	360.93	
Section 4	140	360	322.84	329.62	316.26	
			75.48	74.89	44.67	ΔT
Test 6						
Section 3	140	N/A	398.60	406.75	364.98	
Section 4	70	360	356.27	365.67	356.50	
			42.33	41.08	8.48	ΔT
Test 7						
Section 3	140	N/A	395.53	405.28	363.39	
Section 4	0	360	396.53	408.59	407.30	
			-1.00	-3.31	-43.91	ΔT

Figure 9. NIS Vertiflow Assist vs. Vertiflow

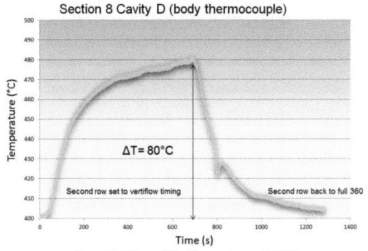

Figure 10. Effects of Vertiflow assist on a QG NIS

CONCLUSIONS

The vertiflow assist cooling system has proven so far to be effective at balancing the vertical and circumferential temperature distribution. By adjusting the timing for both Vertiflow and Vertiflow Assist, we can successfully equalize the temperature from heel to shoulder. The NIS system is currently commercial and has already been installed at several customer facilities. The AIS system is also available and is currently being scheduled for installation at several customer locations. There remains to get a better understanding of the effects of a well-balanced temperature distribution on mold life and repairs.

MULTI-LAYER GLASS THICKNESS MEASUREMENT

Jason Ness, Filipp Ignatovich, Steve Heveron-Smith
Lumetrics, Inc. Rochester, NY

ABSTRACT
Float glass production is associated with challenges in effective and timely quality control. Harsh production environments (high temperatures or caustic chemicals) and non-interruptible manufacturing processes, prevent traditional methods of thickness measurements from detecting manufacturing problems in a timely manner. This results in increased waste, long production duty cycles, and lost revenue. Multilayered glass structures, such as auto and security glass, present additional challenges, as touch gauges or triangulation devices have difficulty measuring individual layers. In recent years, the non-contact method of low-coherence optical interferometry emerged as an effective solution to all of these challenges. Lumetrics has developed and commercialized this non-contact measurement technology, which is now used in the majority of the top glass manufacturers in the world, as well is for other specialty glass processing needs.

INTRODUCTION
Glass is a ubiquitous material which we encounter on a daily basis, whether it's a home window, car windshield, or a smartphone screen. Human ability to make glass products has been known for five thousand years, but traditional flat glass manufacturing was only developed 150 years ago. Since then, flat glass has gone through tremendous improvements, and found applications in all aspects of modern life – from high rise buildings, to aeronautics, to consumer goods. With these improvements came demanding requirements to measure, monitor, and control the glass manufacturing processes.

Architecture glass has gotten thicker while maintaining defect free characteristics. Multiple layers and coatings are used for heat insulation and strength. The automotive industry pushes for thinner glass in order to reduce weight. In communications and consumer goods with multi-layer structures, precise thicknesses are critical to the functionality of the product. Tremendous manufacturing cost savings can be achieved by avoiding the overages that are traditionally built into these products.

In order to effectively control thickness of layers during the manufacturing processes, thickness measurements should be performed in real-time during the production of the glass. Conventional mechanical metrology methods require destruction of the glass after production is done. It increases waste and adds significant cost, when a large batch of product has to be discarded because the problem was not detected in a timely manner. The interferometric method, described in this article, allows for online 100% inspection in real time.

TECHNOLOGY
In low-coherence interferometry [1], light from a source is split into two portions. These two portions form both a sample and reference interferometric arm. The sample portion of the light illuminates the layer stack of a touchscreen, a windshield, or some other device. Each surface within the layer stack reflects back the incident light. The reflected light is then combined with the reference portion of the light. The combined light creates an interferometric pattern.

For a low-coherent light source, the interferometric pattern is visible only under certain conditions, when the path traveled by the light in the sample arm of the interferometer (to and from the corresponding sample interface) is equal to the path in the reference arm of the interferometer. By varying the length of the reference arm of the interferometer, and by measuring the magnitude of the change between the locations where interferometric pattern appears, one can extract the distances between the reflective interfaces within the sample, i.e. the layer thicknesses.

Lumetrics has developed a technology where the interferometer is formed using a communication-grade optical fiber. The path-length change is accomplished by stretching optical fiber using piezo-electric elements. Thus, the distance measurements are accomplished at high speed, with high precision and with reliable long-term continuous operation.

Figure 1 shows an example of the signal acquired for a single layer of glass material, during a single scan of the interferometer. This graph is the equivalent of plotting the reflectivity of the sample versus depth. The peaks indicate the locations of the surfaces of the glass slide, with the first peak corresponding to the top surface of the slide, and the second peak corresponding to the bottom surface of the slide (assuming the slide is lying flat, with the incident light coming from the top). The distance between the two peaks is equal to the optical thickness of the layer. The physical thickness of the layer can be obtained by dividing the optical thickness by the material's refractive index. It is also possible to measure the precise refractive index at the same time as the physical thickness, in a special configuration, where a mirror is placed on the opposite side of the sample.

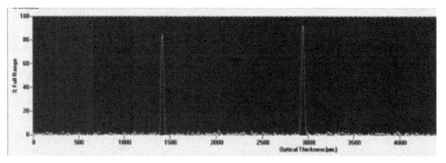

Figure 1. Optical signal obtained for a glass slide. The distance between
the signal peaks is equal to the optical thickness of the slide.

FLOAT GLASS

Measuring glass both at the hot and cold ends is critical to plant efficiency because it allows for the fastest feedback to the plant operators. Low-coherence interferometry can be used to measure glass thickness at multiple locations on the line. Because this technology uses fiber optic cables to connect the sensing probe to the actual measurement system, the customer can position their measurement system thousands of feet away from the oven, where it is unaffected by the heat. The probe is the size of a magic marker and can be mounted in a fixed location or on a scanner, and can be encapsulated into a heat-resistant housing. Multiple probes can also be connected to the same measurement system.

Having sub-micron measurement accuracy, low-coherence interferometry provides users with the ability to track their production and bring glass thickness to tolerances that have not been achieved before. The use of low-coherence interferometry helps plants tune the production parameters faster, and provides them with an effective tool to maintain these parameters for consistent production.

Low-coherence interferometry is well suited for both the online and offline inspection. Customers are using this technology with scanners and manual fixtures to test thickness variability, coating thickness, multi-layer structures, and other features of their glass while in the production process, as well as after. Coated glass, water glass, multi-layer composites can all be measured with any thickness from 0.025mm to 35mm. Figure 2 shows a typical plant layout demonstrating potential locations on either the hot or cold end for measurement.

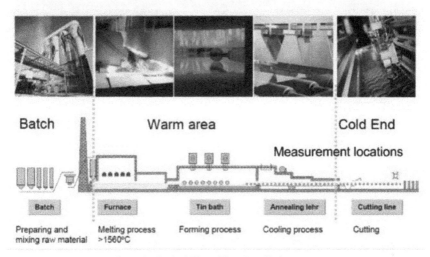

Figure 2. Typical Float Glass Installation

AUTOMOTIVE GLASS

Laminated glass manufacturers are facing the increased need to measure individual layers within the multi-layer glass stacks. For example, in the case of the automotive windshields, manufacturers need to accurately measure the thickness of the Polyvinyl Butyral (PVB) inner-layer. This need arises from the new competitive applications that the automotive manufacturers are designing. For example, the PVB layer can be tailored to accommodate heads-up display functionality, or integrate an acoustic layer to reduce ambient noise. Hand calipers cannot provide these measurements on the production line, or off-line after production, nor can other non-destructive technologies, such as laser triangulation.

Figure 3 shows the layer information from a system screenshot with clear delineation among the glass and different layers of the PVB including the acoustic layer.

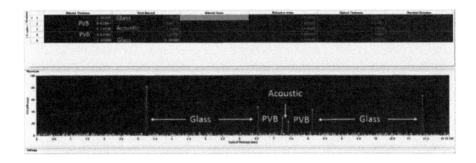

Figure 3. Windshield with acoustic layer and PVB

TOUCH SCREEN STACK

Multi-layer glass panels such as those used in tablet and cell phones are growing in demand. Today, it is difficult to imagine not having a smartphone, tablet or other touchscreen device as part of our daily lives. With the high pace of change in the consumer electronics market, there is continued pressure to further improve the technology associated with these devices. Low-coherence interferometry can be used to generate a visual representation of a cross section of the screen stack. There are a variety of touchscreen technologies, such as resistive, capacitive, optical, acoustic etc. Figure 4 shows a depiction of the layers within the capacitive touchscreen, which is based on detecting the change in the capacitance between two arrays of electrodes, when the human finger approaches the surface of the screen. The touch module consists of two layers of electrodes encompassing an insulating layer (e.g. layer of glass), which acts as a spacer for the capacitor. The module is placed on top of the LCD stack, and the overall system is protected by a cover glass (e.g. Gorilla glass). There are also other variations of the LCD-touchscreen combinations, where touch-enabling layers are integrated with the LCD stack – the on-cell and in-cell technologies.

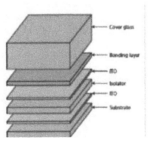

Figure 4: Typical Layers within a touchscreen

To attach different components of the touchscreen together, manufacturers use layers of adhesive. Unlike the thickness uniformity of the glass substrates, the uniformity of the adhesive layer is difficult to control. In the past, the non-uniformity of the layers was acceptable to touch-screen manufacturers. However, as the technology of the LCD touchscreens accelerates, manufacturers are experiencing more and more pressure to ensure that the layer stacks are uniform, to improve image quality as well as the longevity and the durability of touchscreens.

Figure 5 shows the interferometric signal acquired for a smartphone touchscreen [2]. The multitude of peaks indicate numerous layers present under the surface – one can see the top cover glass, the substrates containing ITO electrodes, the spacer between the electrodes, and the adhesive layer between the touch module and the display. Below the adhesive layer (toward the right portion of the graph) are the polarizers and the layer containing thin film transistors (TFT). Some of the layers, such as the ITO and TFT, are much thinner than what the low-coherence interferometer is able to measure. In this case, the two interfaces of such thin layer appear as a single peak, and the measured thickness of the adjacent layers can therefore be slightly larger than the actual material.

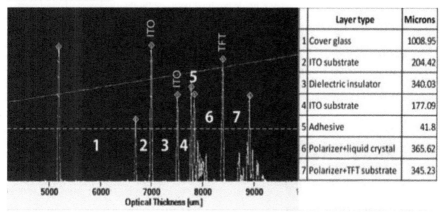

	Layer type	Microns
1	Cover glass	1008.95
2	ITO substrate	204.42
3	Dielectric insulator	340.03
4	ITO substrate	177.09
5	Adhesive	41.8
6	Polarizer+liquid crystal	365.62
7	Polarizer+TFT substrate	345.23

Figure 5. Example of an optical signal obtained for touchscreen. The table on the right show the corresponding thickness

Figure 6 shows measurement results for four different touchscreens from the same manufacturing line. The numbers indicate measurement locations. The measurements immediately show that the thickness of the adhesive layer varies between the different screens, and that the thickness of the adhesive in the lower right corner of the screen, with the exception of the screen 4, is consistently smaller than in the other corners. The film in the middle of the screen is always consistently thicker in all screens but 4. These differences indicate consistent manufacturing problems. The screens were selected from different batches, manufactured under different manufacturing parameters. The parameters appear to be optimal in case of screen 4.

Pt	Sample 1		Sample 2		Sample 3		Sample 4	
	Glass	Adhesive	Glass	Adhesive	Glass	Adhesive	Glass	Adhesive
1	1141.83	187.13	1139.74	120.43	1140.85	153.66	734.09	202.33
2	1134.12	173.84	1141.92	183.61	1146.16	197.37	734.40	204.85
3	1142.34	181.45	1139.64	118.68	1141.17	193.22	737.54	204.85
4	1135.58	105.06	1141.33	86.39	1146.27	166.73	737.77	201.39
5	1140.56	251.10	1139.21	252.29	1146.23	222.43	735.90	202.99

Figure 6. Table illustrating Inconsistency in the adhesive layer thickness

The layer thicknesses in the touchscreens can also be measured within a continuous scan, instead of a single point. To obtain thickness variation along the scan, the screen (or alternatively the optical probe) is moved along the predetermined path, while the thickness of the layers is being continuously measured. The speed of the scan is defined by the measurement rate of the instrument and the required data point density. For example, at the rate of 100 measurements per second, and the scan speed of 100 mm per second, the resulting density of the thickness measurements is 1 measurement per mm. Figure 8 shows the thickness variations of an adhesive layer, acquired during a scan along the right edge of the screen. The graph shows that the adhesive layer thickness changes by over 100% along the scan path. The scan identifies a significant inconsistency in the manufacturing process that cannot be seen with a naked eye, and cannot be directly measured by any mechanical means.

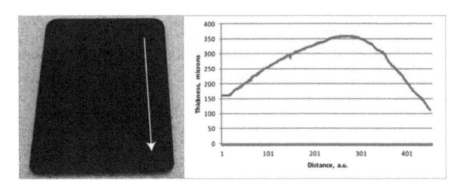

Figure 7. Graph illustrating inconsistency in the adhesive layer thickness.

CONCLUSION

Low-coherence interferometry provides glass manufacturers the ability to gain insights on existing products and processes that are crucial to reduce scrap and improve quality in high volume production. This non-contact technology is superior over other standard techniques because it gives the manufacturer real-time visibility of their production lines, while developing new solutions that meet increased demands of the end user. Lumetrics continues to improve this technology by incorporating enhancements to address complex measurement challenges of manufacturers.

REFERENCES

1. Mart Brezinski, "Optical Coherence Tomography: Principles and Applications", Academic Press, 2006.
2. Measurement of Film Stacks Using White Light Interferometry. (Ignatovich, Gibson, Solpietro, Spaeth, Cotton), AIMCAL / SPE 2014 Proceedings.

THE QUALIFICATION OF A NEW GLASS STRENGTHENING PROCESS

Steven Brown
Bucher Emhart Glass Inc.
Windsor, CT USA

Dubravko Stuhne
Vetroconsult Ltd.
Bülach, Switzerland

ABSTRACT
Thermal strengthening of flat glass has been used for over 80 years in the architectural and automotive industries but it is a new concept for the beverage and container markets. This strengthening process changes the structure and the nature of the glass, adding a stored energy component, which requires that the process be both qualified and verified prior to industry acceptance to show that the end product meets the design specifications in terms of function, strength, durability and safety. By definition, Process Validation = Process Qualification + Process Verification. The Qualification is the evaluation of all the machines used in the production and inspection processes to produce the strengthened containers. The verification phase is a pre-defined production run to verify that the end product (the strengthened bottles) meets all the specifications and requirements. This paper will focus on the efforts required to qualify a new process and to verify the end product for use in the glass marketplace.

INTRODUCTION
In the spring of 2013, we installed a new glass strengthening machine, designed by Bucher Emhart Glass, into a pilot facility owned by Vetropack in Pöchlarn, Austria. The machine is capable of producing thermally strengthened container glass at a rate of 50 bpm. The resulting bottles have an engineered stress profile through the glass that enables increases in strength over annealed glass as a result of residual compressive stresses on both internal and external surfaces.[1]

This new process starts with inspected, annealed glass containers and reheats the articles to approximately 650 °C followed by a rapid and even cooling to approximately 350 °C on all surfaces of the glass, including the inside surface. The end result is a container that has an average of 50MPa of compression on all surfaces. The resulting toughened glass is stronger than annealed glass because glass practically only fails in the presence of tension (together with a critical defect) and all external loads result in maximum tensile stresses on the surfaces. Based on the principle of superposition, this residual compression must first be overcome before the glass goes into tension giving the containers an added margin of usable strength.[2]

The objective of this strengthening program has not only been to increase strength, but to do so while also reducing weight. In fact, one goal was to convert a light weight one-way container into one that is suitable for the returnable market which traditionally has heavier weights for any given capacity.

The article that was chosen for these tests is an amber colored beer bottle having a filled capacity of 330 mL and weighing 200 grams (a typical one way bottle weight for this capacity) to replace an annealed refillable bottle weighing 300 grams. The 300 gram returnable bottle was produced by the blow and blow process while the strengthened bottle was produced by the NNPB process followed with the post process strengthening treatment. The two different bottle designs were also produced by different production plants.

PROCESS VALIDATION

The glass industry is a fairly conservative group making it cumbersome to introduce new products into the marketplace; especially one that changes the foundational principles on which the container market is based, e.g. annealing. The question that we needed to answer was "how do we ensure that the strengthened glass produced by this new process is safe and reliable for use in the returnable consumer marketplace?" This, as it turns out, is not an easy question to answer.

The first step was to qualify every machine that was used in the production and inspection processes. This requires that each machine be certified, calibrated and has the necessary documentation. We then conducted a Failure Modes and Effect Analysis (FMEA) which identified the critical failure modes and resulted in the required paperwork (operating procedures) and corrective actions to mitigate the potential for critical failures. The next step was the actual verification process which required that three consecutive 8 hour days of production be monitored and that samples, at a rate of 5%, be evaluated with the following tests: burst strength, impact strength, drop strength, resistance to thermal shock, cold end coating integrity and surface abrasion (a process that was tailored for this particular application). In total, 21 pallets containing 53,000 bottles were packed.

SAMPLING TEST RESULTS

The actual verification process is ongoing, so the following test results were all gathered during a series of quality assessment tests which were completed in June 2015.

INTERNAL PRESSURE

By way of comparison, a minimum of 300 bottles were burst tested in accordance with EN ISO 7458, in both the as-received and line-simulated (referred to as LS) conditions and for both heavy weight annealed (referred to as ANG) and light weight strengthened containers (referred to as HG). The pre-conditioning line simulation was run for 5 minutes, wet, with filled & capped bottles, at 35 rpm, 25% slip rate and with a bottom abrader in place. As shown in figure 1, the 200 gram HG containers have comparable burst strengths to the 300 gram ANG containers in the as-received condition and, more importantly, the HG containers preserve their strength better than the ANG containers after being subjected to the line simulation pretreatment. In fact, four of the heavy weight annealed bottles failed below the specification limit of 12 bars while none of the light weight strengthened bottles failed. The annealed 200 gram results are included for academic purposes.

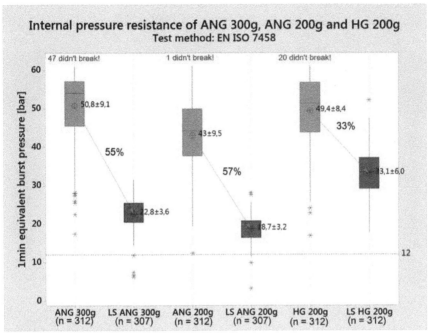

Figure 1. Burst Test Results

IMPACT STRENGTH

The lower contact point sampling tests were evaluated using a pendulum impact tester in accordance with DIN 52295. As was the case with the burst results, the impact strengths of the 200 gram HG containers were higher than the 300 gram ANG containers and the impact strengths were preserved better than the ANG bottles following a line simulation. The results are shown in figure 2. Results were also comparable for the upper contact point.

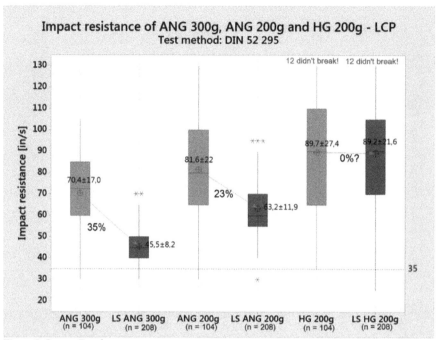

Figure 2. Impact Results

DROP TEST RESULTS

208 empty samples of each of the three bottle designs were dropped from a height of 42 inches (107 cm) onto their bases and their sidewalls. The results are shown in figures 3 (empty bottle drop) and figure 4 (filled drop with 104 samples for each bottle type).

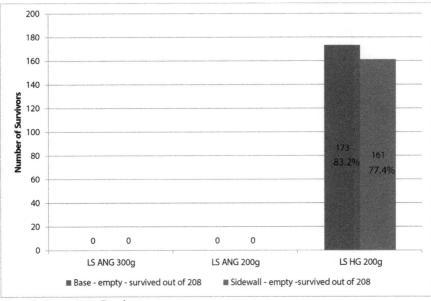

Figure 3. Empty Drop Results

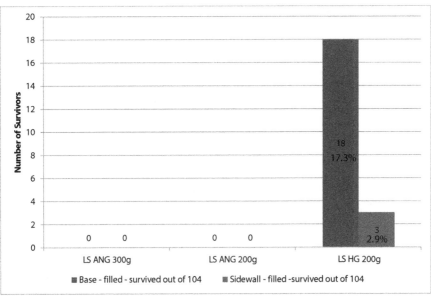

Figure 4. Filled Drop Results

THERMAL SHOCK RESULTS

40 bottles of each type were first exposed to a thermal shock of 90°C that resulted in 7 failures for the 300 gram ANG bottles and 0 for the 200 gram HG bottles. The thermal shock was then increased as shown in table 1.

Table 1 - Thermal Shock Test Results		
	300 gram Annealed (No. of Failures per 40 tests)	200 gram Strengthened (No. of Failures per 40 tests)
Thermal Shock of 90°C	7	0
Thermal Shock of 150°C	40	0
Thermal Shock of 200°C	40	8
Thermal Shock of 250°C	40	33

CONCLUSIONS

The qualification process and the initial quality tests have been completed. A verification run is planned for Nov 2015. If all goes as expected, a sample size of up to 1,000 bottles will be sent to a filler for an actual filling line trial later in Dec 2015. The initial results indicate that the true value of the strengthened containers can be quantified best when the containers are abused (line simulation). After the abuse treatment with the line simulator, the strengthened containers outperformed the 300g and 200g ANG containers in every test (internal pressure resistance, impact resistance, drop resistance etc.) These test results indicate that the strengthened containers have a great potential for use in the returnable glass market. The preferred approach to introduce a new glass container, produced by a new process, into the returnable market is to run a process validation with the help of an outside auditing firm together with an actual Filling line test.

REFERENCES:

1. S. Brown, "Hard Glass – Thermal Strengthening of Container Glass", 73rd Conference on Glass Problems, pages 119 – 130.
2. K. Bratton, S. Brown, D. Stuhne and T. Ringuette, "Hard Glass – Commercial Progress of Thermally Strengthened Container Glass", 75th Conference on Glass Problems, pages 55 – 66.

Author Index